Slaveya Petrova

# Paleo-ecological reconstruction of Perunika village region (Bulgaria)

AF154058

Slaveya Petrova

# Paleo-ecological reconstruction of Perunika village region (Bulgaria)

## Molluscs (Mollusca: Gastropoda, Bivalvia) as indicators for the paleoenvironment

LAP LAMBERT Academic Publishing

**Impressum / Imprint**

Bibliografische Information der Deutschen Nationalbibliothek: Die Deutsche Nationalbibliothek verzeichnet diese Publikation in der Deutschen Nationalbibliografie; detaillierte bibliografische Daten sind im Internet über http://dnb.d-nb.de abrufbar.
Alle in diesem Buch genannten Marken und Produktnamen unterliegen warenzeichen-, marken- oder patentrechtlichem Schutz bzw. sind Warenzeichen oder eingetragene Warenzeichen der jeweiligen Inhaber. Die Wiedergabe von Marken, Produktnamen, Gebrauchsnamen, Handelsnamen, Warenbezeichnungen u.s.w. in diesem Werk berechtigt auch ohne besondere Kennzeichnung nicht zu der Annahme, dass solche Namen im Sinne der Warenzeichen- und Markenschutzgesetzgebung als frei zu betrachten wären und daher von jedermann benutzt werden dürften.

Bibliographic information published by the Deutsche Nationalbibliothek: The Deutsche Nationalbibliothek lists this publication in the Deutsche Nationalbibliografie; detailed bibliographic data are available in the Internet at http://dnb.d-nb.de.
Any brand names and product names mentioned in this book are subject to trademark, brand or patent protection and are trademarks or registered trademarks of their respective holders. The use of brand names, product names, common names, trade names, product descriptions etc. even without a particular marking in this works is in no way to be construed to mean that such names may be regarded as unrestricted in respect of trademark and brand protection legislation and could thus be used by anyone.

Coverbild / Cover image: www.ingimage.com

Verlag / Publisher:
LAP LAMBERT Academic Publishing
ist ein Imprint der / is a trademark of
OmniScriptum GmbH & Co. KG
Heinrich-Böcking-Str. 6-8, 66121 Saarbrücken, Deutschland / Germany
Email: info@lap-publishing.com

Herstellung: siehe letzte Seite /
Printed at: see last page
**ISBN: 978-3-659-28268-3**

# PALEO-ECOLOGICAL RECONSTRUCTION OF PERUNIKA VILLAGE REGION (BULGARIA)

## Molluscs (Mollusca: Gastropoda, Bivalvia) as indicators for the paleoenvironment

*Slaveya Tencheva Petrova, PhD*

*Department of Ecology and Environmental Conservation, Faculty of Biology, University of Plovdiv "Paisii Hilendarski", 24 Tzar Assen Str., Plovdiv, 4000, Bulgaria*
*e-mail: sl.petrova@abv.bg*

## TABLE OF CONTENTS

# I.   INTRODUCTION

*WHAT IS PALEONTOLOGY?*

Paleontology is the study of the biota of past geologic ages and the relationships in its development since appearing billions of years ago to the modern era. Term "paleontology" comes from the Greek words: *palaios* – old; *ontos* - being, creature; *logos* – study.

As material for research in paleontology serve fossilized remains of ancient plants and animals, named fossils (from Latin *fossilis* - to dig), buried in the layers of the Earth's crust. These are a variety of shells, skeletons or parts of them, imprints of the bodies of plants and animals, as well as traces of their live activities. Importance of fossils of ancient organisms is enormous - they are the foundation of the most important methods for determining the relative age of rocks and correlation of cross sections from different areas. Furthermore, most of the organic remains are primary indicators of environmental conditions of habitat. Third, but not least, the fossils are the only material evidence of the long history of life on Earth and the basis on the restoration of the laws of biological evolution.

By its nature and subject, the paleontology is a biological discipline and therefore many scientists often named it as paleobiology. Different sections of paleontology are the mirror image of the main sections of biology. Some of these sections examine various organism groups, while others are subjected to certain peculiarities in the structure and function of organisms, their relationships with the environment, their distribution and so on:

*Invertebrate Paleontology* and *Vertebrate Paleontology* - object of study are all phyla of animals, and they develop knowledge of the morphology, systematics and evolution of animals.

*Paleobotany* - object of study are the plants, algae and fungi.

*Paleoecology*: Study of the ecosystems of the past, the relationships between the ancient organism and the environment they lived.

Paleontology incorporates knowledge from biology, geology, ecology, anthropology, archaeology, and even computer science to understand the processes that have led to genesis and eventual destruction of the different types of organisms since life arose.

3

Paleoecology is the study of "the relationships between organisms in ancient times and their environment" (Gekker, 1957). Object of study of paleoecology are habitat conditions and the way of life of the organisms in past geological periods, the relationships between organisms and their environment (abiotic and biotic), the organisms changes in the life cycles on Earth, and the paleoecological aspects of taphonomy changes in the processes of fossilization. Aim of paleoecology is therefore to build the most detailed model possible of the life environment of previously living organisms found today as fossils.

Paleoecology and ecology deal with many general and specific concepts, among which are "way of life" and "habitat". According Gekker (1957) the term "way of life" includes: diet, locomotion, breeding, defense and attack, etc. "Habitat" means the environment in which an organism lives. Largest areas of habitation are aquatic and terrestrial. Within these main areas are separated some environments with specific characteristics. In the marine area such environments are the beach, tidal zone, shelf and deep water and the open (pelagic) area of the seas. In the continental could separate land, river, lake, pond, lagoon environment and some others.

Main unit of paleoecology is the paleobiocenosis. To this system belong organic remains (fossils), traces and products of life activity of ancient organisms and concerning rocks. Paleoecological research aims to identify the components of paleoecosystem, the interactions between them, their structure and functions.

Analysis of the "ancient organisms - environment" is only possible after reconstruction of the paleoecosystem. Such reconstruction takes into consideration the complex interactions among environmental factors such as temperature, food supplies, and degree of solar radiation. Fossil record is rich in biological and ecological information, but the quality of this information is uneven and incomplete. Often much of this information is lost or distorted by the fossilization process or diagenesis of the enclosing sediments, making interpretation difficult.

Studies of numerous fossil deposits from a geological period, located in different regions provide useful information for the reconstruction of the paleogeographic and paleoecological conditions in different stages of Earth's history.

Interpretation of paleoecological data requires a working knowledge of biology and involves the use of substantive uniformitarianism, analogy, and parsimony.

Geologists have learned that Earth's systems have changed since the Archean. Archean-Proterozoic transition is believed to mark the beginning of a more stable planet. Concept of substantive uniformitarianism, which is based on an understanding that the materials, conditions, and rates of processes have remained relatively constant through time, has played a part in the interpretation of paleoeocological data. This idea is applicable to strata deposited since the Late Proterozoic, when metaphytes and metazoans first evolved.

Analogy (or actuopaleontology) involves the application of modern organismic features to ancient organisms. This principle may be applied to individuals (with regard to form and function), community structure (species diversity, organizational and trophic structure), and population dynamics (response to time-independent environmental factors).

Parsimony involves the use of the simplest, or most parsimonious, explanation to decipher the data. That is, the explanation that uses the fewest steps, beginning with the cause, through the intermediate causes, responses, and effects to the final result, is the most desirable when interpreting the information. Parsimony is not restricted to paleoecological studies or paleontological studies in general. Parsimony is a central tenet of all scientific inquiry.

Ecological and paleoecological studies can focus on an individual species (known as autecology) or on many species (known as synecology). Autecological investigations commonly are concerned with the organism's response to its environment: the morphological adaptations that the organism has evolved in order to meet the minimum requirements for survival, the organism's behavioral traits acquired to most efficiently exploit its environment, or the impact of the environment on the individual. Some aspects of fossil populations are difficult to evaluate because of the nature of the fossil record. Because of this, autecological studies have focused on the structure and evolution of fossil populations rather than on population characters that may help interpret paleoenvironments. Attributes of fossil populations provide information that reflects organismic adaptions for survival within the abiotic (physical) and biotic (biological) constraints imposed during their geologic history (Gastaldo et al., 1996).

5

## II.   GEOCHRONOLOGY

*CENOZOIC ERA*

Cenozoic Era is the most recent of the three major subdivisions of animal history (the other two are the Mesozoic and Paleozoic Eras). Cenozoic (65.5 million years ago to present) is divided into three periods: the Paleogene (65.5 to 23.03 million years ago), Neogene (23.03 to 2.6 million years ago) and the Quaternary (2.6 million years ago to present). Paleogene and Neogene are relatively new terms that now replace the deprecated term, Tertiary.

Paleogene is subdivided into three epochs: Paleocene (65.5 to 55.8 million years ago), Eocene (55.8 to 33.9 million years ago), and Oligocene (33.9 to 23.03 million years ago). Neogene is subdivided into two epochs: Miocene (23.03 to 5.332 million years ago) and Pliocene (5.332 to 2.588 million years ago) (Dates from the International Commission on Stratigraphy's International Stratigraphic Chart, 2013).

*EOCENE EPOCH*

Eocene epoch lasted from about 55.8 to 33.9 million years ago. Early Eocene (Ypresian) is thought to have had the highest mean annual temperatures of the entire Cenozoic era, with temperatures about 30°C; relatively low temperature gradients from pole to pole; and high precipitation in a world that was essentially ice-free. Land connections existed between Antarctica and Australia, between North America and Europe through Greenland, and probably between North America and Asia through the Bering Strait. It was an important time of plate boundary rearrangement, in which the patterns of spreading centers and transform faults were changed, causing significant effects on oceanic and atmospheric circulation and temperature.

In the Middle Eocene, the separation of Antarctica and Australia created a deep water passage between those two continents. This changed oceanic circulation patterns and global heat transport, resulting in a global cooling event observed at the end of the Eocene. By the Late Eocene, the new ocean circulation resulted in a significantly lower mean annual temperature, with greater variability and seasonality worldwide. Lower temperatures and increased seasonality drove increased body size of mammals, and caused a shift towards increasingly open savanna-like vegetation, with a corresponding reduction in forests.

At the beginning of the Cenozoic (66 million years ago) Rhodopes were one relatively consolidated massif. During Paleogene this massif disintegrated into few smaller ridges. At its eastern part the so called Central, South-Eastern and Harmanli blocks were separated (Ivanov 1961, Boyanov 1971). Volcano-tectonic depressions were found between the crystalline blocks. They are characterized by active volcanism and sedimentation. Around these faults initially formed graben, shallow and disconnected with each other continental and saltwater steep shores. Gradually these pools expanded, gone deeper and have consolidated in one.

In the Priabonian, the area of this pool has continued to increase, the bottom gradually sinking. In the shallow, warm and relatively large marine pool takes place rich sedimentation. In the Southern and South-Eastern part of the depression the so called Metlichky coral reef was formed. It occupied an area of about 30 km between the Chorbadzhyisko village, the Elbasan River valley, the region of Krumovgrad town, South and East of the Irantepe Peak and reached the Perunika village. It is made of thick, massive limestone with foraminifers' shells, coral, algae, bivalves, gastropods and others. Its width reaches up to 6 km, and the thickness of up to 90 m (Georgiev, 2002).

At the end of the Eocene (about 56-34 million years), on the territory of present Bulgaria there were few separate salt water basins with some large islands within, which were presented by the contemporary massifs of the Sredna Gora, Strandzha, Kraishte, Stara Planina mountains and the central mountain ridge of the Rhodopes. Water of these seas has a normal halinity and the regional climate was tropical one (Yordanov, 1962; Boshev et al., 1966).

Paleontological studies of the Paleogene in the Eastern Rhodopes (region of Kardzhali) are presented in the publications of Karagyuleva (1964) and Sapundjieva (1964), which described the fossil representatives of the classes Anthozoa, Bivalvia, Gastropoda, Echinoidea, and Foraminifera. In the 70s and 80s the same authors have developed detailed descriptions of lithostratigraphy of the region during the Eocene-Oligocene based on the established fossil fauna.

Paleoecological studies of the geological evolution of the Eastern Rhodopes are still missing; the properties and structure of ancient communities in the region remained unexplored.

### III.  EXPLORING THE FOSSIL TROVE

Fossil material was collected during 2011-2013 in the region of Perunika village, Eastern Rhodopes, Bulgaria (Fig.1), from the soil surface (down to 20 cm) and accounted more than 800 fossil specimens. Fossils were stored in plastic bags (Petrova et al., 2012). In laboratory conditions they were cleaned, sorted and identified to the lowest possible taxon, using Bulgarian and International systematic guide books and a reference collection. Systematics, distribution and ecology of molluscs are presented according to the references in the International paleodatabase (www.paleodb.org).

Measurement of fossils was carried out with a digital caliper. Abbreviations in measurements mean: L – shell length; W – shell width; H – shell height.

Collected fossils are stored in the Department of Ecology and Environmental Conservation, Faculty of Biology, University of Plovdiv "Paisii Hilendarski", Bulgaria.

Software for paleontological statistical analysis "PAST" ver. 2.07 (Hammer et al., 2001) was used in calculating the index range of Simpson, Shannon index, Margalef index, index of Berger-Parker, for the purposes of the paleoecological analyses.

**Fig.1.** Map of Europe and location of the fossil trove

## IV. BIOSTRATIGRAPHIC CHARACTERISTICS OF THE FOSSIL TROVE

Firstly, as the basis of all paleontological research, it is needed to explore the geology and stratigraphy of fossil trove location area. Detailed and proper geological information include data on the type of minerals and rocks in the area, their structure and formation. Stratigraphic study concerns the identification of the sequence of formation of all types of rocks in this district, as well as the age and spatial relationships between them.

Next step is to check the tectonic structure and tectonic development of the area. Major structural forms are folds (anticlines and syncline), over thrusts, thrusts and faults, caused by the movement of the earth's crust under the influence of internal earth forces.

### GEOLOGICAL CHARACTERISTICS OF THE AREA

Geology of Bulgaria (and in particular that of the Rhodopes) cannot be seen divorced from the geology of the surrounding land and the whole crust, as in the country have developed the same geological processes, as they settled in the Balkan Peninsula, in Europe, part of the Asian and the African continent, which lands are in close geographical and geological context (Boyanov & Goranov, 2001).

According to the latest findings on the geological structure of the Earth, the Earth crust is divided into platform districts and geosynclines that constantly changing over geological history. Bulgaria and the lands of the Balkan Peninsula belong mainly to the Mediterranean geosynclines, which is occupied by the folds of the Alpine-Himalayan system, formed by the youngest Alpine cycle.

Before finally reach its present shape, the area of the Alpine-Himalayan (Mediterranean) geosynclines repeatedly and in different periods of the Mesozoic and Cenozoic was flooded by the waters of various seas. Abiotic conditions in these seas were favorable for organisms. Thus a rich fauna was developed, abundant fossilized remains of which are found in sedimentary rocks of almost all formations.

Beginning of the collapse of the Rhodopes and the emergence of volcano-tectonic depression between fragments can be placed between Maastricht-Dan (Goranov, 1960). Priabonian in the Eastern Rhodopes is divided into four lithological horizons, namely clastic breccia, carboniferous-sandstone, limestone and volcanic-sedimentary (Boyanov et al., 1963).

Breccia-conglomerate horizon has limited distribution. It usually fills snatch formed among crystalline. Compositions of the pieces correspond to rocks of the pad - a variety of gneisses, amphibolites and marbles. Sorting is not observed. Bulk is generally sandy and less sandy clay or lime-sand. Thickness ranges from a few meters to over 170 meters. Transition with top-lying carboniferous sandstone horizon is a gradual but quick.

Carboniferous-sandstone formation consists of two layers. Lower one is not well developed and is represented by alternating sandstones, siltstones, marls, small gravel conglomerates and untenable thin layers of coal and carbonaceous shale. Upper part is significantly more widespread and is deposited throughout the basin. It is consisted of irregularly alternating with each other sandstones, conglomerates, siltstones and marls. Thickness of the layer ranges from a few meters to over 160 m.

Marl-limestone series was put the normal and rapid lithological transition on carboniferous-sandstone formation. Transition takes place from 2-3 m thick alternations of calcareous sandstone, sandy limestone and limestone. In the southern and southeastern part of the depression is formed dusters coastal reef. After the deposition of marl-limestone layer, a new transgression of the sea basin has taken place. Process is accompanied by underwater volcanic activity (Georgiev, 2002).

In summary, probably during Maastrichtian, some small and isolated pools filled with coarse material were formed in the area. In Priabonian, as a result of transgression, the clastic breccia horizon is postponed. Isolated pools gradually extended deepening and merge with each other until a single, relatively large body of water. Breccia-conglomerate, carboniferous-sandstone and limestone-marl horizons were formed super-positionally, followed by the volcanogenic-sedimentary. Simultaneously, the early appearances of intense collisional volcanism were marked.

Stratigraphic structure of the Rhodope Mountain, its lithological and facial features, and the expression of magmatism are closely related and dependent on its tectonic structure and tectonic development.

In tectonic aspect Bulgaria is related to the tectonic evolution of the Alpine system or the Alpine orogen, formed on the site of the Western Mediterranean geosynclines, occupied by the so-called Tethys basin. In the South-Eastern part of the European continent, the Alpine orogen is represented by the mountains of the Balkan Peninsula.

In its current formation, Rhodope massif occupies the Southern and South-Western parts of the country. It is limited by the number of fault systems, which are considered as tectonic boundaries of the massif. So, on the North is limited to the Maritsa fault zone, on the West - with Struma fault zone, south of its border runs along the coastline of the Aegean (Aegean) Sea and on the East borders of the East fault (Fig.2).

In the Rhodopes are released several structural stages made of different age metamorphic and sedimentary-volcanogenic complexes, formed during the tectonic processes. These structural levels can be summarized in two clearly divergent structural complexes: inferior, old - Precambrian and upper - young, built mainly of sedimentary and volcanic rocks of Priabonian, Oligocene and Pliocene.

Finally, as regards the tectonic evolution of the Rhodope Mountains, it can be noted that the Rhodope region consists mainly of two complex rocks - old (probably Precambrian and Paleozoic) and young (Paleogenic).

Tectonics has also manifested in two stages - an old, during which they were wrinkled old rocks, and younger (Alpine), which has affected the younger formations. Following the first tectonic phase are formed the basic structure of the Rhodope region, and during the second tectonic phase are formed rifts in synclinal areas, folds and thrusts.

Thick rock series that built the base of the Rhodope Mountain were layered before the Cambrian, when all the Rila-Rhodope area was under water. At the end of the Precambrian and the beginning of the Paleozoic the large anticlinal structures are formed. This was accompanied

11

with the introduction of large quantities of granite magma. By deep cracks that magma is injected as spinous and among rocks, leading to their metamorphism.

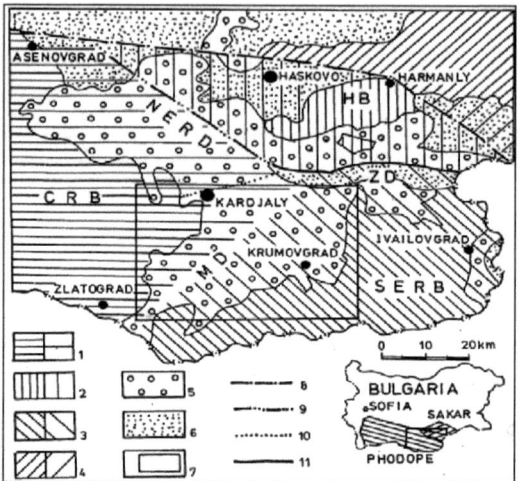

Fig.2. Schematic overview of the Eastern Rhodopes (in Jovchev et al., 1971, with additions by Georgiev, 2002)

Legend: 1 - Central Rhodope block (CRB); 2 - Harmanliyan block (HB); 3 – South-Eastern Rhodopean block (SERB); 4 - Sakar unit; 5 - Paleogene sediments and volcanics; 6 - Neogene-Quaternary sediments; 7 - border research area; 8 - Maritsa fault zone; 9 - Pchelarovski faults beam; 10 - Ardino dislocation; 11 - Zlatoustovska dislocation; NERD – North-Eastern depression; MD - Momchilgrad depression; ZD - Zlatoustovska depression.

In the early Palaeozoic the Rhodope region became land. This land has existed throughout the Mesozoic and early Cenozoic. Upper Eocene (Priabonian) was a turning point in the geological evolution of the Rhodope region. It began to sink, thus allowing water to be spilled in the Eastern and Central part of the Rhodopes. Decrease was accompanied by vertical faulting on land, manifested mainly in the synclinal areas.

After Priabonian, the territory of Rhodope Massif was subjected to the third phase of the Pyrenees cycle. The fold, however, is manifested only in the synclinal areas. Shaped old anticlines were not affected. Tectonic stress was come from the north, causing the dip of the folds in the south. Multiple faults are formed simultaneously with the creation of different types of folds and tectonic structures.

In the Late Neogene the Rhodope region has been subjected to very complex earth movements. Eastern and Southern parts began to sink, and Western and Middle have risen. This has led to new cracking of the crust in Pliocene and to the formation of valleys filled with fresh water. Aegean Sea has detached at that time. Rise of the Rhodope region continued, leading to drainage waters from Pliocene lakes and to the formation of high mountain ridges (Ivanov, 1960).

## V. SYSTEMATICS OF THE PHYLUM MOLLUSCA

Molluscs are one of the most diverse groups of animals on the planet, with at least 50,000 living species (and more likely around 200,000). They are a clade of organisms with soft bodies which typically have a "head" and a "foot" region. Often their bodies are covered by a hard exoskeleton, as in the shells of snails and clams or the plates of chitons (Bunje, 2003).

Parts of almost every ecosystem in the world, molluscs are extremely important members of many ecological communities. They range in distribution from terrestrial mountain tops to the hot vents and cold seeps of the deep sea, and range in size from 20-meter-long giant squid to microscopic aplacophorans, a millimeter or less in length, that live between sand grains.

Phylum Mollusca is divided into seven classes (Fig.3):

1)   GASTROPODA (snails, slugs, limpets, sea hares)

2)   BIVALVIA (scallops, clams, mussels, etc.)

3)   APLACOPHORA (spicule-covered, worm-like animals)

4)   MONOPLACOPHORA (limpet-like "living fossils")

5)   POLYPLACOPHORA (chitons)

6)   SCAPHOPODA (tusk shells)

7)   CEPHALOPODA (nautilus, squids, octopuses, ammonites)

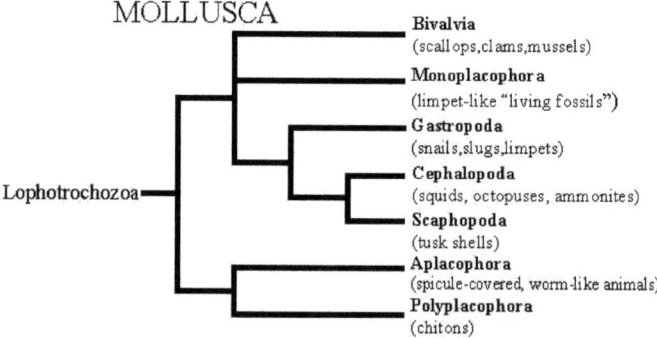

**Fig.3.** Molluscan systematics

Gastropods are one of the most diverse groups of animals, both in form, habit and habitat. They are by far the largest group of molluscs, with more than 62,000 described living species, and they comprise about 80% of living molluscs. They have a long and rich fossil record from the Early Cambrian that shows periodic extinctions of sub clades, followed by diversification of new groups.

Gastropods have figured prominently in paleobiological and biological studies, and have served as study organisms in numerous evolutionary, biomechanical, ecological, physiological, and behavioral investigations. They are extremely diverse in size, body and shell morphology, and habits and occupy the widest range of ecological niches of all molluscs.

Gastropods live in every conceivable habitat on Earth. They occupy all marine habitats ranging from the deepest ocean basins to the supralittoral, as well as freshwater habitats, and other inland aquatic habitats including salt lakes. They are also the only terrestrial molluscs, being found in virtually all habitats ranging from high mountains to deserts and rainforest, and from the tropics to high latitudes.

Gastropod feeding habits are extremely varied, although most species make use of a radula in some aspect of their feeding behavior. They include grazers, browsers, suspension feeders, scavengers, detritivores, and carnivores. Carnivory in some taxa may simply involve grazing on colonial animals, while others engage in hunting their prey. Some gastropod carnivores drill holes in their shelled prey, this method of entry having been acquired independently in several groups, as is also the case with carnivory itself. Some gastropods feed suctorially and have lost the radula (Bunje, 2003).

## SHORT CHARACTERISTICS OF CLASS BIVALVIA

The second most diverse group of molluscs behind gastropods, bivalves are one of the most important members of most marine and freshwater ecosystems. In fact, there are well over 10,000 described species of bivalve, found from the deepest depths of the oceans, to the smaller streams.

Bivalves are easily recognized by their two-halved shell. They can burrow into the sediment or live on the ocean floor. Some can even move around through the water by snapping their shell open and shut to swim. Not all bivalves still have a shell though; some have evolved a reduced shell or have completely lost the shell.

Most bivalves are filter feeders, using their gills to capture particulate food such as phytoplankton from the water. Food is bound in mucus that is carried by cilia along food grooves on the edges of the gills to the mouth region. Here particles are sorted on the ciliated labial palps before they enter the mouth. Food then moves to the stomach, which is large and complex with sophisticated ciliary sorting mechanisms, and usually, a rotating hyaline rod which liberates enzymes into the stomach. Digestion is carried out in the large paired digestive diverticula (the stomach).

First occurrences of Bivalvia are found in Lower Cambrian deposits, but it is not until the Lower Ordovician that bivalve diversification, both taxonomic and ecological, explodes in the fossil record. This diversification continues unabated through the Phanerozoic, with relatively small losses at the end-Permian and end-Cretaceous extinction events. From relatively humble beginnings, they have diversified and expanded to become dominant members of most marine ecosystems (Waller, 1998).

# VI. PALEO-ECOLOGICAL ANALYSES OF THE FOSSIL TROVE NEAR PERUNIKA VILLAGE (EASTERN RHODOPES)

*TAXONOMYCAL CHARACTERISTICS*

From all collected fossils, 654 specimens (more than 80%) were identified as molluscs - 618 gastropods and 36 mussels. Class Gastropoda was presented by at least 25 species from 21 genera (Table 1), and class Bivalvia – with 10 genera and at least 11 species (Table 2). Many corals and echinoids were found also, but they are not included in this analysis.

**Table 1.** Characteristics of identified gastropod fossils from Perinuka village (Eastern Rhodopes)

| № | Taxon | Fossil number | Locomotion | Life habit | Diet | References |
|---|---|---|---|---|---|---|
| 1. | *Velates perversus* | 28 | Actively mobile | Epifaunal | Omnivore-Grazer | Kiessling, 2004 |
| 2. | *Tympanotonus trochleare diaboli* | 52 | Actively mobile | Epifaunal | Grazer | Kiessling, 2004; Hendy et al., 2009 |
| 3. | *Chondroceritium filigrana* | 7 | Actively mobile | Epifaunal | Herbivore | Kiessling, 2004; Abbott and Dance, 1986 |
| 4. | *Lyria decora* | 6 | Actively mobile | Epifaunal | Carnivore | Kiessling, 2004; Hendy et al., 2009 |
| 5. | *Rimella multiplicata* | 21 | Actively mobile | Epifaunal | Omnivore | Kiessling, 2004; Hendy et al., 2009 |
| 6. | *Cirsotrema bourdoty* | 39 | Slow moving | Low level epifaunal | Carnivore | Beu et al., 1990; Kiessling, 2004 |
| 7. | *Sycostoma bulbiforme* | 5 | Actively mobile | Epifaunal | Carnivore | Kiessling, 2004; Hendy et al., 2009 |
| 8. | *Ficus helveticus* | 4 | Actively mobile | Epifaunal | Carnivore | Kiessling, 2004; Hendy et al., 2009 |
| 9. | *Amaurellina angustata* | 16 | Actively mobile | Epifaunal | Grazer | Kiessling, 2004; Hendy et al., 2009 |
| 10. | *Globularia vapincana* | 25 | Actively mobile | Epifaunal | Grazer | Kiessling, 2004; Hendy et al., 2009 |
| 11. | *Globularia grossa* | 9 | Actively mobile | Epifaunal | Grazer | Kiessling, 2004; Hendy et al., 2009 |
| 12. | *Globularia patula* | 13 | Actively mobile | Epifaunal | Grazer | Kiessling, 2004; Hendy et al., 2009 |
| 13. | *Globularia sigaretina* | 36 | Actively mobile | Infaunal | Carnivore | Kiessling, 2004; Aberhan et al., |

| | | | | | | 2004 |
|---|---|---|---|---|---|---|
| 14. | *Globularia crassatina* | 2 | Actively mobile | Epifaunal | Grazer | Kiessling, 2004; Hendy et al., 2009 |
| 15. | *Globularia sp.* | 37 | - | - | - | - |
| 16. | *Cepatia cepacea* | 17 | Slow moving | Shallow infaunal | Carnivore | Kiessling, 2004 |
| 17. | *Euspira achatensis* | 11 | Actively mobile | Infaunal | Carnivore | Kiessling, 2004; Aberhan et al., 2004 |
| 18. | *Borsonia biplicata* | 32 | Actively mobile | Epifaunal | Carnivore | Kiessling, 2004; Hendy et al., 2009 |
| 19. | *Borsonia sp.* | 5 | - | - | - | - |
| 20. | *Terrebellum belemnitoideum* | 5 | Actively mobile | Epifaunal | Omnivore-Grazer | Kiessling, 2004; Hendy et al., 2009 |
| 21. | *Canarium auriculatum* | 5 | Actively mobile | Epifaunal | Omnivore-Grazer | Abbott and Dance, 1986; Kiessling, 2004; Hendy et al., 2009 |
| 22. | *Charonia vincenti* | 1 | Actively mobile | Epifaunal | Carnivore | Kiessling, 2004 |
| 23. | *Campanile lachesis* | 1 | Actively mobile | Epifaunal | Grazer | Kiessling, 2004; Hendy et al., 2009 |
| 24. | *Cassis harpaeformis* | 2 | Actively mobile | Epifaunal | Carnivore | Kiessling, 2004; Hendy et al., 2009 |
| 25. | *Tectus lucasianus* | 1 | Actively mobile | Epifaunal | Grazer | Kiessling, 2004; Hendy et al., 2009 |
| 26. | *Cryptoconus filosus* | 2 | Actively mobile | Epifaunal | Carnivore | Kiessling, 2004; Hendy et al., 2009 |
| 27. | *Burtinella spirulaea* | 10 | Actively mobile | Epifaunal | - | Kiessling, 2004 |
| 28. | *Gastopoda sp. -* undetermined | 226 | | | - | |
| | **Total Gastropoda** | **618** | | | | |

**Table 2.** Characteristics of identified bivalve fossils from Perunika village (Eastern Rhodopes)

| № | Taxon | Fossil number | Locomotion | Life habit | Diet | References |
|---|---|---|---|---|---|---|
| 1. | *Pecten sp.* | 5 | Faculatively mobile | Epifaunal | Suspension feeder | Aberhan et al., 2004; Kiessling, 2004 |
| 2. | *Macrosolen sp.* | 2 | Faculatively mobile | Infaunal | Suspension feeder | Aberhan et al., 2004 |
| 3. | *Crassatella sp.* | 3 | Faculatively mobile | Infaunal | Suspension feeder | Eames, 1951; Aberhan et al., 2004 |
| 4. | *Cardim gratum* | 1 | Faculatively mobile | Infaunal | Suspension feeder | Todd, 2001; Aberhan et al., 2004 |
| 5. | *Lima rara* | 4 | Faculatively mobile/Attached | Epifaunal | Suspension feeder | Sanchez Roig, 1926; Aberhan et al., 2004; Kiessling, 2004 |
| 6. | *Chlamys rhodopianus* | 1 | Stationary attached | Epifaunal | Suspension feeder | Kiessling, 2004 |
| 7. | *Chlamys sp.* | 3 | - | - | - | - |
| 8. | *Pycnodonte brongniarti* | 1 | Stationary attached | Epifaunal | Suspension feeder | Steuber, 2002; Aberhan et al., 2004; Kiessling, 2004 |
| 9. | *Pitar bonnetensis* | 3 | Faculatively mobile | Infaunal | Suspension feeder | Aberhan et al., 2004; Kiessling, 2004; Hendy et al., 2009 |
| 10. | *Pitar villanovae* | 1 | Faculatively mobile | Infaunal | Suspension feeder | Aberhan et al., 2004; Kiessling, 2004; Hendy et al., 2009 |
| 11. | *Tellina decorata ovalina* | 1 | Faculatively mobile | Deep infaunal | Deposit feeder | Aberhan et al., 2004; Kiessling, 2004 |
| 12. | *Barbatia appendiculata sokolovi* | 1 | Stationary attached | Epifaunal | Suspension feeder | Aberhan et al., 2004; Mikkelsen and Bieler, 2008 |
| 13. | *Barbatia sp.* | 1 | - | - | - | - |
| 14. | *Bivalvia* indet. | 9 | - | | | |
| | **Total Bivalvia** | **36** | | | | |

Phyllum **MOLLUSCA**

Class **Gastropoda**

Subclass **Orthogastropoda** Ponder and Lindberg, 1997

Superorder **Vetigastropoda** Salvini-Plawen, 1980

Superfamily **Trochoidea** (Rafinesque, 1815)

Family **Trochidae** Rafinesque, 1815

Subfamily **Trochinae** Rafinesque, 1815

Genus **Tectus** de Montfort, 1810

*Tectus lucasianus* (Brongniart, 1823)

**Material:** 1 fragmented specimen

**Dimensions:** L = 18.55 mm; W = 19.34 mm; H = 6.25 mm

**Description:** High conical shell, composed of a low, spreading down whirls, separated by deep narrow seam. Decoration of the shell consists of two spiral rows buds.

**Deposits in Bulgaria:** near the town of Ivaylovgrad.

**Level in Bulgaria:** Oligocene.

**Total stratigraphic and geographic distribution:** Upper Eocene of Macedonia and Southern France; Lower and Middle Oligocene of Vicentino (Fig.4).

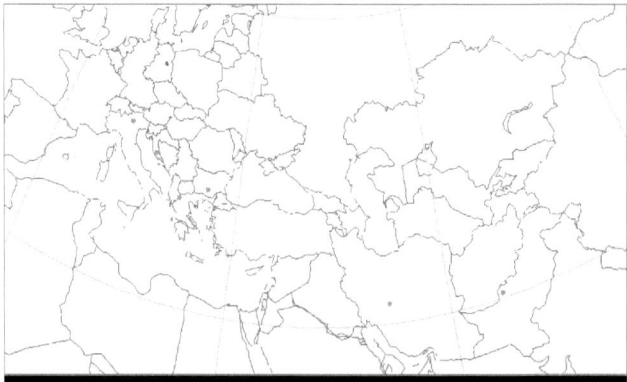

**Fig.4.** Distribution map of *Tectus sp.*

Subclass **Orthogastropoda** Ponder and Lindberg, 1997

Superorder **Neritopsina** Cox, 1960

Superfamily **Neritoidea** (Rafinesque, 1815)

Family **Neritidae** Rafinesque, 1815

Subfamily **Velatinae** Bandel, 2001

Genus *Velates* de Montfort, 1810

*Velates perversus* (Gmelin, 1791)

**Material:** 28 specimens (19 well preserved and 9 fragmented)

**Dimensions:** L = 22.36-48.54 mm; W = 19.54-39.03 mm; H = 9.58-20.25 mm

**Description:** Shell is oval rounded, highly conical, elongated backward. Beak is located in the front third of the length of the shell, it is not large, blunt, turn right and down. Decoration of shell is very typical and preserved in almost all specimens - it consists of fine, closely spaced elliptical lines of growth. Incorrect lighter or darker colored stripes are distinguished (Fig.5).

**Deposits in Bulgaria:** Lower and Middle Eocene near city of Varna; Priabonian near the village of Kraymorie (Burgas), Haskovo, Kardzhali; Oligocene near the town of Kardzhali.

**Level in Bulgaria:** Lower Eocene-Oligocene.

**Total stratigraphic and geographic distribution:** Paleocene of the Paris Basin, Iran, Egypt, Priabonian of Spain, French Alps, Northern Italy, the Bavarian Alps, Switzerland, Austria, Hungary, Northern Transylvania, Ukraine, Georgia, Armenia, Herzegovina, Dalmatia, Albania , Macedonia, Sicily, Egypt, Sahara, Somalia, Arabia and Central Asia (Fig.6).

**Fig.5.** *Velates perversus*

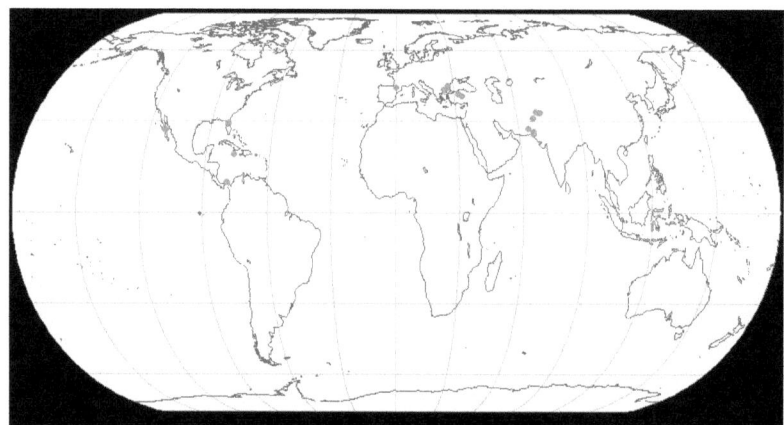

**Fig.6.** Distribution map of *Velates sp.*

Subclass **Prosobranchia** Milne-Edwards, 1848

Superorder **Caenogastropoda** Cox, 1959

Order **Neotaenioglossa** Haller, 1882

Superfamily **Cerithiacea**

Family **Vermetidae** d'Orbigny, 1840

Genus **Burtinella** Mörch, 1861

***Burtinella spirulaea*** (Lamarck)

**Material:** 10 specimens (6 well preserved and 4 fragmented)

**Dimensions:** L = 10.32-22.97 mm; W = 7.54-11.58 mm; H = 5.05-7.33 mm

**Description:** Shell is flat. Whirls are separated by a pronounced seam. Inner part is convex, while the outer is narrower and ends with a blunt edge. Shell surface is decorated with lines of growth, which are slightly wavy curved.

**Deposits in Bulgaria**: Varna, Kotel, Belogradchik, Aytos.

**Level in Bulgaria**: Ypresian-Priabonian.

**Total stratigraphic and geographic distribution**: Ypresian of the Southern Alps; Lutetian of the Southern Alps, Bavarian Alps, Crimea; Priabonian of the Bavarian Alps, Switzerland, Northern Italy, Armenia (Fig.7).

**Fig.7.** Distribution map of *Burtinella sp.*

Subclass **Prosobranchia** Milne-Edwards, 1848

Superorder **Caenogastropoda** Cox, 1959

Order **Sorbeoconcha** Ponder and Lindberg, 1997

Family **Potamididae** H. Adams & A. Adams, 1854

Genus *Tympanotonos* Klein, 1753 in Schumacher, 1817

Subgenus *Tympanotonos* s. str. Klein, 1753 in Schumacher, 1817

*Tympanotonus trochleare diaboli* (Brongniart, 1823)

**Material:** 52 specimens (16 well preserved и 36 fragmented)

**Dimensions:** L = 8.46-13.92 mm; W = 7.21-12.78 mm; H = 17.05-32.93 mm

**Description:** Shell is thin, high conical, composed of 10-12 slightly bulging whirls, separated by a deep suture. Decoration consists of two main spiral ribs disposed on upper and lower ends of the bend, provided with bumps. Typically, the upper flange is embossed with larger spots. In the bulk of our specimens between the two ribs is inserted a third, not so in relief and with small papules. Buds of the spiral ribs are located one above the other and are connected together by narrow transverse ribs, which give a view of the threaded shell. Cells are rectangular, square and even rhombic. In lower site of the last turn has a torn spiral ridge, behind which there are 7-8 spiral ribs crossed by fine thick hair lines (Fig.8).

**Deposits in Bulgaria**: Dolno Lukovo, Ivaylovgrad; Krumovgrad.

**Level in Bulgaria**: Priabonian.

**Total stratigraphic and geographic distribution**: Priabonian of Southern France, Northern Transylvania, Hungary and Dalmatia (Fig.9).

**Fig.8.** *Tympanotonus trochleare diaboli*

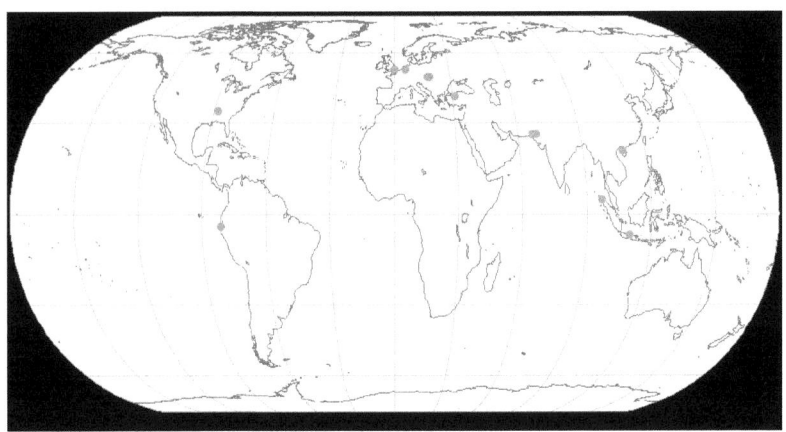

**Fig.9.** Distribution map of *Tympanotonus sp.*

Subclass **Prosobranchia** Milne-Edwards, 1848

Superorder **Caenogastropoda** Cox, 1959

Order **Sorbeoconcha** Ponder and Lindberg, 1997

Family **Cerithiidae** Fleming, 1828

Genus *Cerithium* Bruguiare, 1789

Subgenus *Chondrocerithium* Monterosato in Cossmann, 1906

*Chondrocerithium filigrana* (v. Koenen, 1891)


**Material:** 7 specimens (3 well preserved and 4 fragmented)

**Dimensions:** L = 8.47-11.9 mm; W = 7.12-10.4 mm; H = 20.1-27 mm

**Description:** Middle whirls are weakly convex, almost flat. Last one is great and ends with a short channel. Sculpture of the shell consists of helical fins covered with longitudinally elongated buds. On the middle whirls there are observed three coarse spiral ribs. One is located below the top seam, the other - above the bottom seam, and the third one is located closer to the bottom. On the last whirl the number of these ribs is four. In addition, there are some secondary ribs between two principal. There are also many soft spiral ribs of third order (Fig.10).

**Deposits in Bulgaria**: near the town of Ivaylovgrad.

**Level in Bulgaria**: Priabonian.

**Total stratigraphic and geographic distribution**: Upper Eocene of North Germany (Fig.11).

**Fig.10.** *Chondrocerithium filigrana*

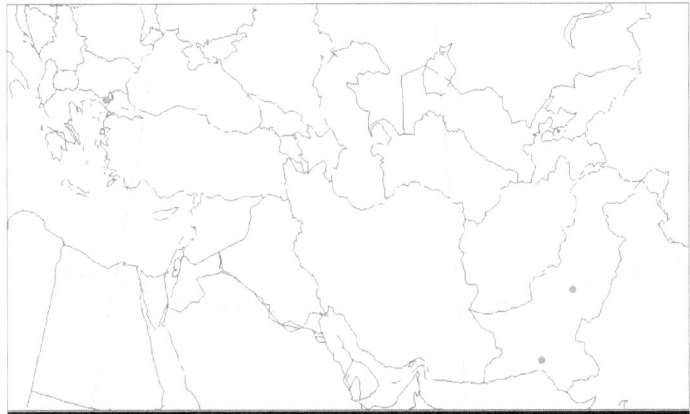

**Fig.11.** Distribution map of *Chondrocerithium sp.*

Subclass **Prosobranchia** Milne-Edwards, 1848

Superorder **Caenogastropoda** Cox, 1959

Order **Ptenoglossa** (Gray, 1853)

Superfamily **Epitonoidea** Berry, 1812

Family **Epitoniidae** Berry, 1910

Genus *Cirsotrema* Mörch, 1852

*Cirsotrema bourdoty* (Boury, 1883)


**Material:** 39 specimens (31 well preserved and 8 fragmented)

**Dimensions:** L = 12.75-24.41 mm; W = 8.33-21.27 mm; H = 21.2-44.87 mm

**Description**: Shell is high conical, composed of 7 convex whirls. Seam is deep, somewhat hidden. Ornamentation of the shell is composed of 13-14 lamellar transverse ribs, located slightly biased. On the first whirl the ribs are smooth, but the middle and lower ribs are with visible longitudinal stripes. Some of the ribs are thickened, especially in the last turn. At the top of the transverse ribs there are thin, not very well developed spines. Intercostal areas are slightly wider than the ribs with delicate spiral lines of three orders of magnitude. Most preferably are those, expressed by the first order, and arranged at equal intervals. Between any two of them, have one more thin, and among them - one or two of the third order. At the bottom of the last whirl there is a spiral rib as a cordon (Fig.12).

**Deposits in Bulgaria:** near the town of Provadia.

**Level in Bulgaria**: Upper Eocene.

**Total stratigraphic and geographic distribution:** Middle Eocene of Paris Basin (Fig.13).

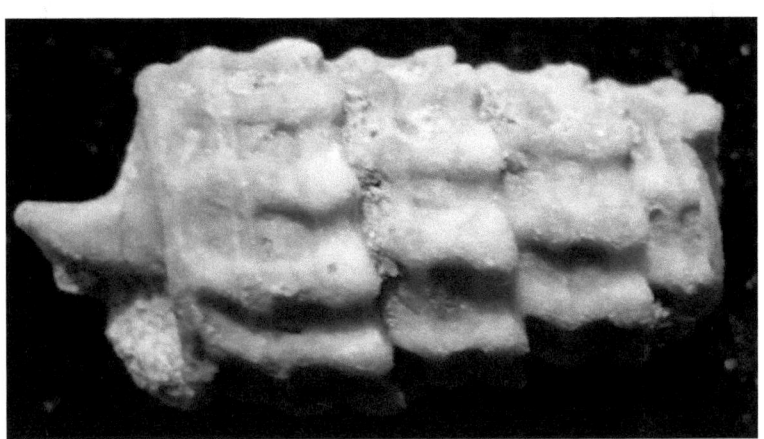

**Fig.12.** *Cirsotrema bourdoty*

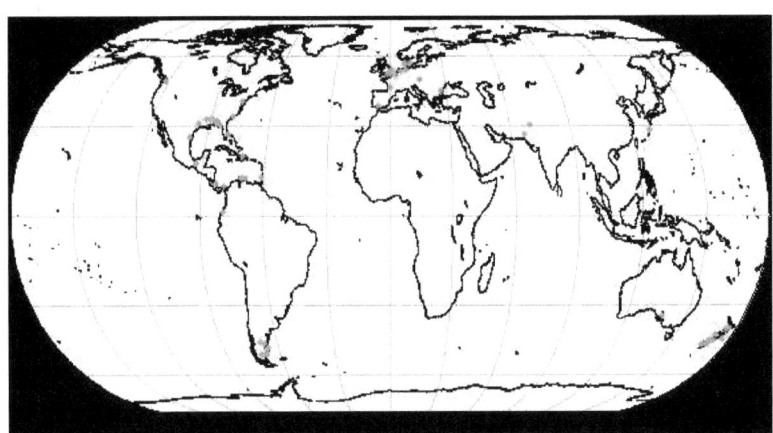

**Fig.13.** Distribution map of *Cirsotrema sp.*

Subclass **Prosobranchia** Milne-Edwards, 1848

Superorder **Caenogastropoda** Cox, 1959

Order **Neogastropoda** Thiele, 1929

Superfamily **Buccinoidea** Rafinesque, 1815

Family **Melongenidae** Gill, 1867

Subfamily **Melongeninae** (Gill, 1871)

Genus *Sycostoma* Cox, 1931

*Sycostoma bulbiforme* (Lamarck, 1803)


**Material:** 5 fragmented specimens

**Dimensions:** L = 36.78-74.75 mm; W = 41.44-60.38 mm; H = 32.56-105 mm

**Description**: Shell is oval, elongated and consists of 6 whirls. Middle ones are slightly rounded and properly expand down. Last whirl is significantly large and occupies almost four fifths of the total height of the shell. At the top it is considerably swollen, narrowed down and ends with a short siphon, which in our forms is not retained. Seam is narrow and shallow. Shell is smooth with fine lines of growth (Fig.14).

**Deposits in Bulgaria**: near the town of Aytos.

**Level in Bulgaria:** Priabonian.

**Total stratigraphic and geographic distribution**: Ypresian of the Paris Basin; Lutetian of England, the Paris Basin, Italy; Bartonian of England, Paris Basin, France, Belgium (Fig.15).

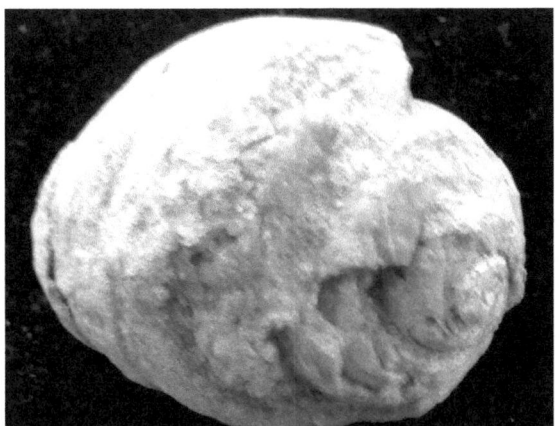

**Fig.14.** *Sycostoma bulbiforme*

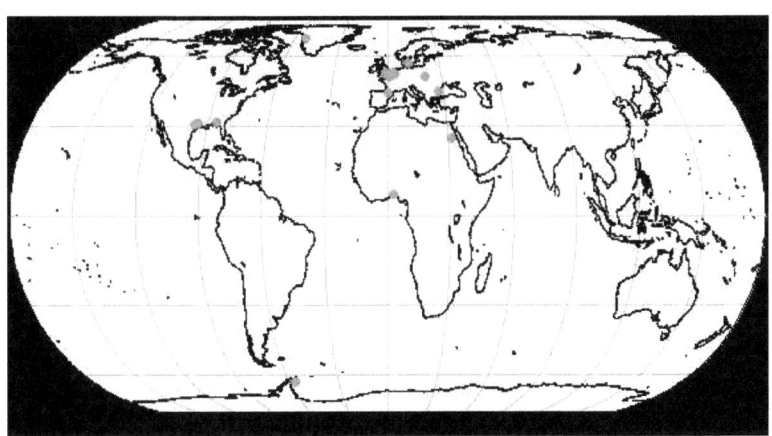

**Fig.15.** Distribution map of *Sycostoma sp.*

Subclass **Prosobranchia** Milne-Edwards, 1848

Superorder **Caenogastropoda** Cox, 1959

Order **Neogastropoda** Thiele, 1929

Superfamily **Muricoidea** de Costa, 1776

Family **Volutidae** Rafinesque, 1815

Subfamily **Volutinae** Rafinesque, 1815

Genus *Lyria* Gray, 1847

*Lyria decora* (Beyrich, 1853)

**Material:** 6 fragmented specimens

**Dimensions:** L = 14.56-18.85 mm; W = 12.4-16.12 mm; H = 22.7-39.29 mm

**Description**: Shell is extended with spindle shape. It is composed of 8 slightly protuberant whirls, of which the latter occupies more than half of the total height. Seam is clearly expressed, shallow. Last whirl is slightly narrower at the bottom. Whirls are decorated with 16-18 rounded transverse ribs. They are smooth and thickened. On the last turn they are slightly curved back to the siphon. Spiral ribs on the bottom of the last turn due to leaching can be seen very weak (Fig.16).

**Deposits in Bulgaria**: near the town of Aytos.

**Level in Bulgaria**: Priabonian.

**Total stratigraphic and geographic distribution**: Middle and Upper Eocene of England; Upper Eocene of Ukraine and Northern Germany (Fig.17).

**Fig.16.** *Lyria decora*

**Fig.17.** Distribution map of *Lyria sp.*

Subclass **Prosobranchia** Milne-Edwards, 1848

Superorder **Hypsogastropoda** (Ponder and Lindberg, 1997)

Family **Strombidae** Rafinesque, 1815

Genus *Rimella* Röding, 1798

*Rimella multiplicata* (Bellardi, 1852)


**Material:** 21 specimens (8 well preserved and 13 fragmented)

**Dimensions:** L = 6.52-10.05 mm; W = 6.17-9.35 mm; H = 16.9-28.9 mm

**Description**: Shell is high conical, composed of 7-8 flat, slowly expanding downward whirls, separated by a pronounced seam. Last one is slightly swollen. Shell is decorated by transverse ribs, separated by wider intercostal. They get lost in the middle of the last turn. On the bottom of the last whirl are visible several thin closely spaced spiral ribs (Fig.18).

**Deposits in Bulgaria:** near the town of Parvomay.

**Level in Bulgaria:** Priabonian.

**Total stratigraphic and geographic distribution**: Priabonian of Vicentino (Fig.19).

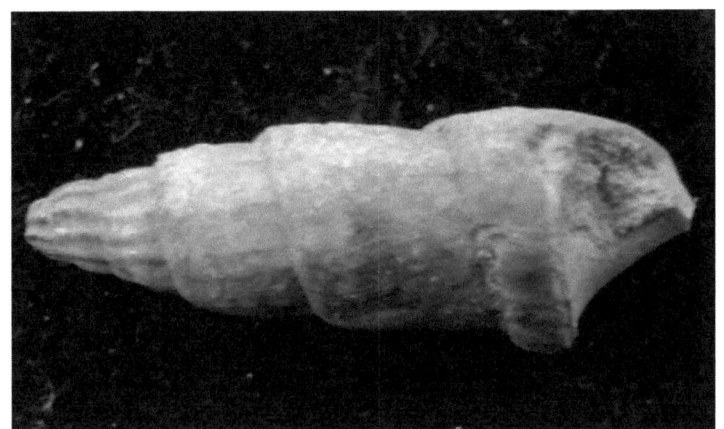

**Fig.18.** *Rimella multiplicata*

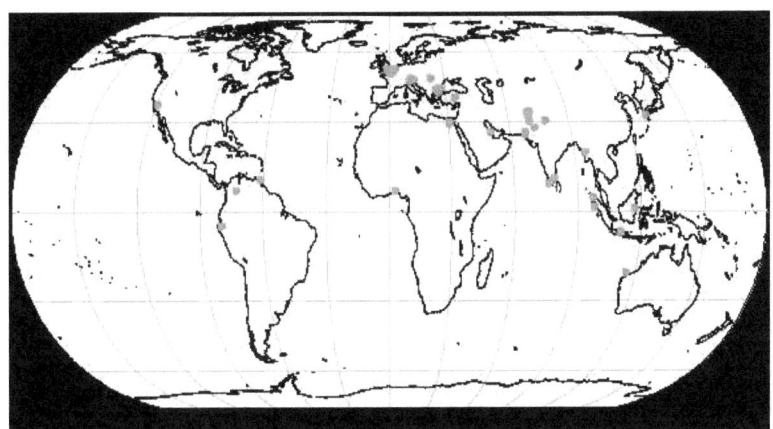

**Fig.19.** Distribution map of *Rimella sp.*

Subclass **Prosobranchia** Milne-Edwards, 1848

Superorder **Hypsogastropoda** (Ponder and Lindberg, 1997)

Family **Strombidae** Rafinesque, 1815

Genus *Strombus* Linnaeus, 1758

Subgenus *Canarium* (Schumacher, 1817)

*Canarium auriculatum*

**Material:** 5 specimens (2 well preserved and 3 fragmented)

**Dimensions:** L = 57.78-68.3 mm; W = 30.97-42.92 mm; H = 53.71-92.81 mm

**Description:** Shell is elongated, composed of 7 whirls, the latter of which occupies 5/6 of the total height. First 6 whirls are low, slightly protuberant. They quickly expand downward to form a low sharp cone. Seam is deep. Shell is almost smooth (Fig.20).

**Deposits in Bulgaria:** Near the town of Burgas, Sliven, Krumovgrad, Haskovo.

**Level in Bulgaria**: Priabonian-Oligocene.

**Total stratigraphic and geographic distribution**: Southern France and Western Alps (Fig.21).

**Fig.20.** *Canarium auriculatum*

**Fig.21.** Distribution map of *Canarium sp.*

Subclass **Prosobranchia** Milne-Edwards, 1848

Superorder **Hypsogastropoda** (Ponder and Lindberg, 1997)

Family **Ficidae** Meek, 1864

Genus *Ficus* Röding, 1798

Subgenus *Fulguroficus* Sacco, 1890

*Ficus helveticus* (Mayer, 1867)


**Material:** 4 fragmented specimens

**Dimensions:** L = 25.77-40.26 mm; W = 30.43-36.16 mm; H = 33.38-40.2 mm

**Description**: Shell has an irregular pear-shaped, with not much extended siphonal channel. Stitching is clear, shallow. Last whirl is large, strong bulging. Surface of the shell is covered with a thin rounded spiral ribs disposed at equal intervals. They are cut by thin transverse ribs. Cellules obtained by the intersection of the helical and transverse ribs are narrow, rectangular, transversely elongate (Fig.22).

**Deposits in Bulgaria**: near the town of Aytos.

**Level in Bulgaria**: Priabonian.

**Total stratigraphic and geographic distribution**: Bartonian of the Swiss Alps; Priabonian of Vicentino, Bavarian Alps, Georgia (Fig.23).

**Fig.22**. *Ficus helveticus*

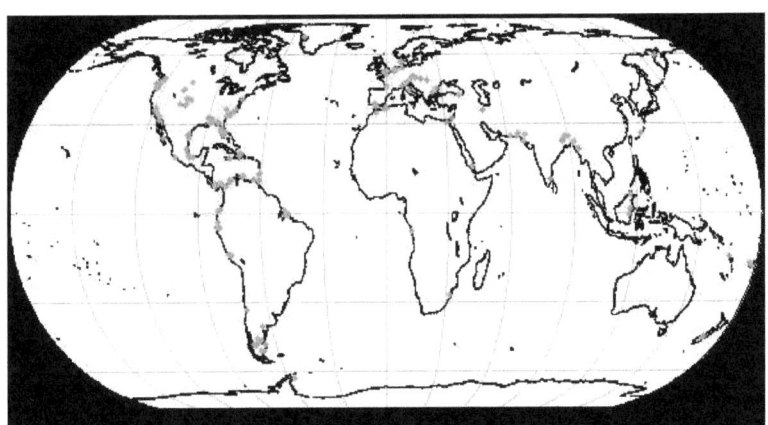

**Fig.23.** Distribution map of *Ficus sp.*

Subclass **Prosobranchia** Milne-Edwards, 1848

Superorder **Caenogastropoda** Cox, 1959

Superfamily **Ampullinoidea** Cossmann, 1918

Family **Ampullinidae** Cossmann, 1918

Subfamily **Globulariinae**

Genus *Amaurellina* Fischer, 1885

Subgenus *Crommium* Cossmann, 1888

*Amaurellina angustata* (Grateloup, 1827)


**Material:** 16 specimens (7 well preserved and 9 fragmented)

**Dimensions:** L = 26.15-52.66 mm; W = 22.04-43.15 mm; H = 34.8-58.63 mm

**Description:** Shell is medium-sized, thick, consisted of 8 whirls. They are not high, rapidly expanding and very slightly protuberant, almost flat. Seam is deep. Last whirl is very large, bulging and occupies nearly ¾ of the total height. Umbilicus is narrow, slightly open. Shell surface is covered with very fine spiral lines. At the bottom they strongly recurve (Fig.24).

**Deposits in Bulgaria**: near the town of Momchilovgrad, Blagoevgrad, Haskovo.

**Level in Bulgaria**: Eocene.

**Total stratigraphic and geographic distribution**: Middle Eocene of Vicentino, Southern France (Fig.25).

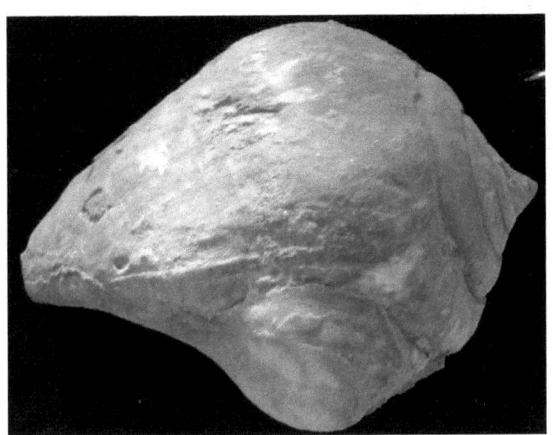

**Fig.24.** *Amaurellina angustata*

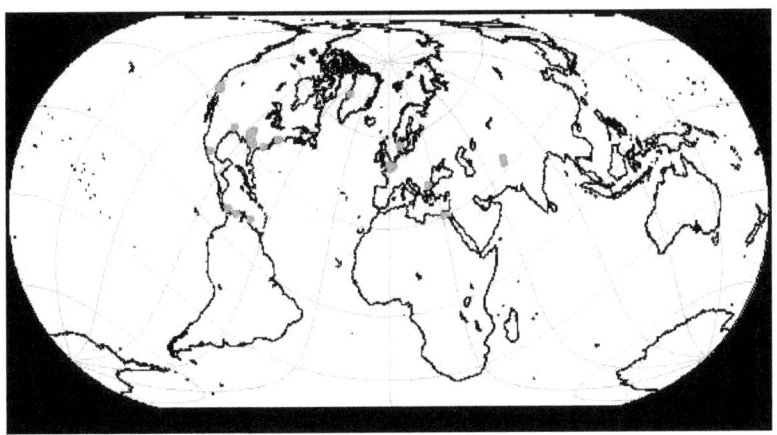

**Fig.25.** Distribution map of *Amaurellina angustata*

Subclass **Prosobranchia** Milne-Edwards, 1848

Superorder **Caenogastropoda** Cox, 1959

Superfamily **Ampullinoidea** Cossmann, 1918

Family **Ampullinidae** Cossmann, 1918

Subfamily **Globulariinae**

Genus *Globularia* Swainson, 1840

Subgenus **Ampullinopsis** Conrad, 1865

*Globularia crassatina* (Lamarck, 1804)

**Material:** 2 specimens

**Dimensions:** L = 33.23-58.51 mm; W = 21.15-46.75 mm; H = 21.43-54.21 mm

**Description**: Shell is solid, consisted of 6-7 whirls. Middle ones are relatively high, very slightly convex, separated by a deep suture. Last whirl is very large and takes up 3/4 of the height of the entire shell. Ornamentation of the shell consists of fine spiral lines, closely spaced. They intersect at an acute angle and form a dense network with rhombic cells.

**Deposits in Bulgaria**: near the town of Haskovo, Kardzhali, Momchilovgrad, Krumovgrad.

**Level in Bulgaria**: Oligocene.

**Total stratigraphic and geographic distribution**: Middle Oligocene of England, the Paris Basin, Southern France, Macedonia (Fig.26).

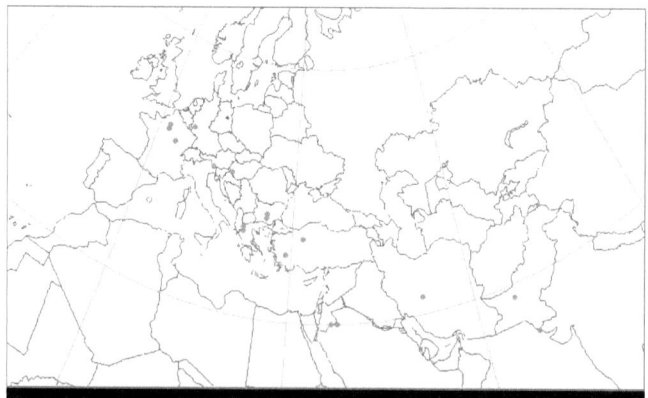

**Fig.26.** Distribution map of *Globularia crassatina*

42

Subclass **Prosobranchia** Milne-Edwards, 1848

Superorder **Caenogastropoda** Cox, 1959

Superfamily **Ampullinoidea** Cossmann, 1918

Family **Ampullinidae** Cossmann, 1918

Subfamily **Globulariinae**

Genus *Globularia* Swainson, 1840

Subgenus *Globularia* s. str. Swainson, 1840

*Globularia patula* (Lamarck, 1804)

**Material:** 13 fragmented specimens

**Dimensions:** L = 16.77-51.18 mm; W = 11.45-26.4 mm; H = 11.42-44.61 mm

**Description:** Shell is thick, composed of 7 whirls. They are short, slightly convex and rapidly expanding down. Last one is significantly large and occupies four fifths of the height of the entire shell. Seam is well defined.

**Deposits in Bulgaria**: near the town of Aytos, Chirpan, Burgas, Veliko Tarnovo, Krumovgrad.

**Level in Bulgaria**: Priabonian.

**Total stratigraphic and geographic distribution**: England, Paris Basin, Belgium, Bavarian Alps, Switzerland, Northern Transylvania, Nice; Upper Eocene of the Paris Basin, Georgia, Northern Transylvania, Ukraine (Fig.27).

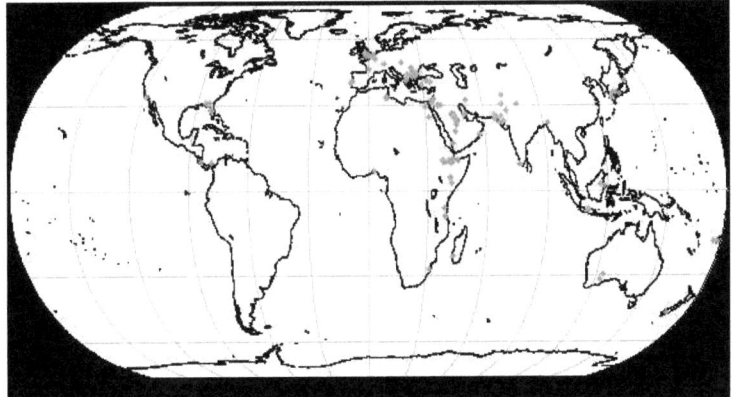

**Fig.27.** Distribution map of *Globularia patula*

43

Subclass **Prosobranchia** Milne-Edwards, 1848

Superorder **Caenogastropoda** Cox, 1959

Superfamily **Ampullinoidea** Cossmann, 1918

Family **Ampullinidae** Cossmann, 1918

Subfamily **Globulariinae**

Genus *Globularia* Swainson, 1840

Subgenus *Globularia* s. str. Swainson, 1840

*Globularia grossa* (Deshayes, 1864)

**Material:** 9 fragmented specimens

**Dimensions:** L = 19.34-37.93 mm; W = 16.26-35.1 mm; H = 26.16-49.82 mm

**Description**: Medium to large individuals, the shell is composed of 6 whirls, which are very slightly swollen. They are separated from each other by well-defined smooth seam. Last whirl is bigger, more bloated and covers 3/4 of the total height of the shell. Decoration of the shell consists of uniformly dense, fibrous transverse ribs, at the bottom strongly curved backwards. They correspond to the lines of growth. Some of them are shown bold and break increases. In well-preserved shell are observed very fine spiral lines.

**Deposits in Bulgaria**: Burgas.

**Level in Bulgaria**: Priabonian.

**Total stratigraphic and geographic distribution**: Middle and Upper Eocene of the Paris Basin and England (Fig.28).

**Fig.28.** Distribution map of *Globularia grossa*

44

Subclass **Prosobranchia** Milne-Edwards, 1848

Superorder **Caenogastropoda** Cox, 1959

Superfamily **Ampullinoidea** Cossmann, 1918

Family **Ampullinidae** Cossmann, 1918

Subfamily **Globulariinae**

Genus *Globularia* Swainson, 1840

Subgenus *Globularia* s. str. Swainson, 1840

*Globularia sigaretina* (Lamarck, 1804)


**Material:** 36 specimens (22 well preserved and 14 fragmented)

**Dimensions:** L = 16.52-49.42 mm; W = 11.97-37.5 mm; H = 18.45-60.21 mm

**Description:** Shell is consisted of 5-6 lower slightly convex whirls. Last one is very big and swollen. Seam is relatively deep and smooth. Surface of the shell is covered with fine lines which are embossed in the end of the last whirl.

**Deposits in Bulgaria:** near the town of Sliven, Ivaylovgrad, Chirpan, Aytos, Tarnovo.

**Level in Bulgaria**: Priabonian.

**Total stratigraphic and geographic distribution:** Lutetian of England, Paris Basin, Belgium, Bavarian Alps, Vicentino, Egypt; Bartonian of England, Paris Basin, Vicentino, Switzerland; Upper Eocene of England, South West France, the Western Alps, Bavarian Alps, Dalmatia, Herzegovina (Fig.29).

**Fig.29.** Distribution map of *Globularia sigaretina*

45

Subclass **Prosobranchia** Milne-Edwards, 1848

Superorder **Caenogastropoda** Cox, 1959

Superfamily **Ampullinoidea** Cossmann, 1918

Family **Ampullinidae** Cossmann, 1918

Subfamily **Globulariinae**

Genus *Globularia* Swainson, 1840

Subgenus *Globularia* s. str. Swainson, 1840

*Globularia vapincana* (d'Orbigny, 1850)


**Material:** 25 specimens (19 well preserved and 6 fragmented)

**Dimensions:** L = 16.06-51.04 mm; W = 12.91-37.1 mm; H = 20.17-59.12 mm

**Description**: Shell is relatively high, consisting of 8 whirls, which are rapidly expanding downwards and are arranged stepwise. Seam is well defined. Last whirl is considerable and covers ¾ of the total height. Decoration of the shell is composed of well-expressed dense growth lines, curved back at the bottom. Some are more clearly expressed and marked a break in growth (Fig.30).

**Deposits in Bulgaria**: near the town of Aytos, Ivaylovgrad, Krumovgrad.

**Level in Bulgaria**: Priabonian

**Total stratigraphic and geographic distribution:** Lower Priabonian of the Western Alps, Switzerland, South-Eastern France, Northern Italy, Northern Transylvania (Fig.31).

46

**Fig.30.** *Globularia vapincana*

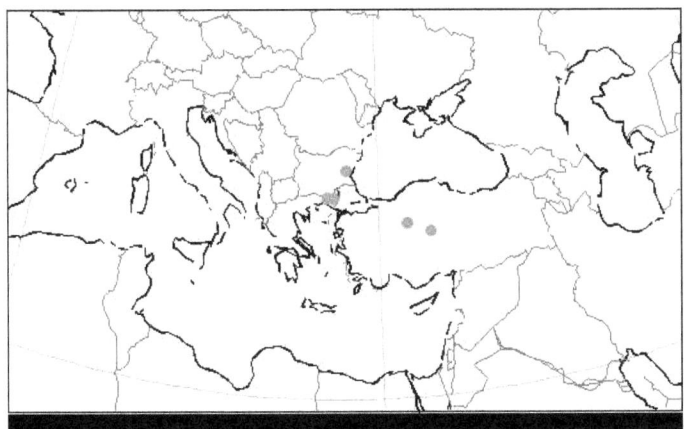

**Fig.31.** Distribution map of *Globularia vapincana*

Subclass **Prosobranchia** Milne-Edwards, 1848

Superorder **Caenogastropoda** Cox, 1959

Superfamily **Naticoidea** Guilding, 1834

Family **Naticidae** Guilding, 1834

Genus *Cepatia* Gray, 1847

*Cepatia cepacea* (Lamarck, 1804)

**Material:** 17 specimens (10 well preserved and 7 fragmented)

**Dimensions:** L= 17.86-61.02 mm; W = 13.97-46.11 mm; H = 10.75-25.77 mm

**Description:** Medium-sized shell, consisting of approximately 7 wide lower whirls, separated from each other with deep seam. Last whirl is considerably big and swollen. Shell surface is covered with fine spiral lines (Fig.32).

**Deposits in Bulgaria**: Burgas, Chirpan, Haskovo, Kardzhali, Krumovgrad.

**Level in Bulgaria:** Priabonian.

**Total stratigraphic and geographic distribution:** Ypresian of Egypt, England; Lutetian of England, France, Italy, Istria, Bavarian Alps, Egypt, Asia Minor; Bartonian of England, France, Italy, Herzegovina, Dalmatia; Priabonian of France, Switzerland, Dalmatia, Hungary, Bavarian Alps, Georgia, Northern Transylvania; Eocene of Pakistan (Fig.33).

**Fig.32.** *Cepatia cepacea*

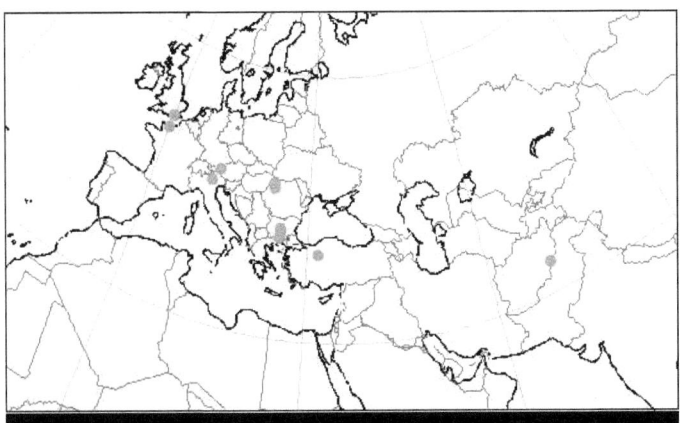

**Fig.33.** Distribution map of *Cepatia cepacea*

Subclass **Prosobranchia** Milne-Edwards, 1848

Superorder **Caenogastropoda** Cox, 1959

Superfamily **Naticoidea** Guilding, 1834

Family **Naticidae** Guilding, 1834Subfamily **Policinae** Gray, 1847

Genus *Euspira* Agassiz, 1838

*Euspira achatensis* (Recluz, 1837)

**Material:** 11 specimens (9 well preserved and 2 fragmented)

**Dimensions:** L = 17.8-28.41 mm; W = 13.48-21.37 mm; H = 18.25-31.83 mm

**Description:** Shell is consisted of 6 whirls, separated from each other with deep narrow seam. Shape of the shell varies - sometimes is spherical, sometimes is elongated. Last whirl is large, bulging and occupies 3/5 of the height of the shell. Decoration of the shell consists only of the fine lines of growth, which are closely spaced. At the lower end of the last whirl they are strongly swept backwards (Fig.34).

**Deposits in Bulgaria:** Aytos.

**Level in Bulgaria:** Priabonian.

**Total stratigraphic and geographic distribution:** Priabonian of the Western Alps, Northern Transylvania, Ukraine, Northern Germany, Belgium; Oligocene of the Paris Basin, Vicentino, Georgia (Fig.35).

**Fig.34.** *Euspira achatensis*

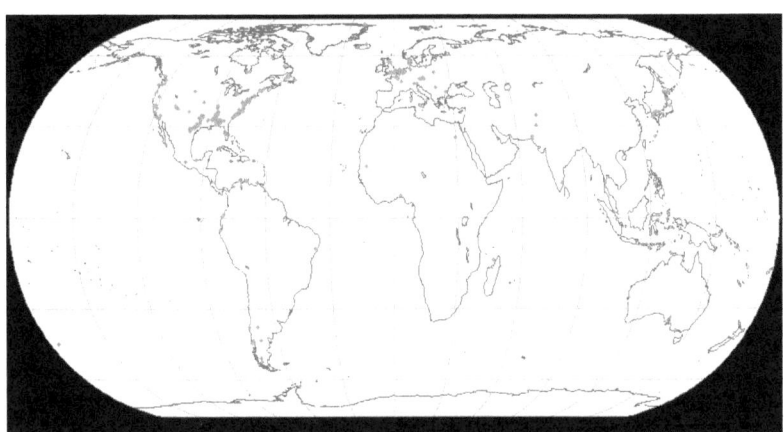

**Fig.35.** Distribution map of *Euspira sp.*

Subclass **Prosobranchia** Milne-Edwards, 1848

Superorder **Caenogastropoda** Cox, 1959

Order **Neogastropoda** Thiele, 1929

Superfamily **Conoidea** Rafinesque, 1815

Family **Conidae** Fleming, 1822

Genus **Cryptoconus** Koenen, 1867

*Cryptoconus filosus* (Lamarck, 1804)

**Material**: 2 specimens

**Dimensions:** L = 19.56-22.76 mm; W = 13.45-15.52 mm; H = 35.21-39.12 mm

**Description:** Shell is fusiform with pointed ends, composed by 7-8 whirls. Stitching is clear, smooth. Whirls are slightly convex. Last one is bulging and occupies half of the total height of the shell. On the upper corners there is a narrow, slightly concave line, separated by a narrow channel. Shell surface is decorated with rounded spiral ribs crossed by fine lines.

**Deposits in Bulgaria:** Kyustendil, Momchilgrad.

**Level in Bulgaria**: Priabonian-Oligocene.

**Total stratigraphic and geographic distribution**: Lutetian of the Paris basin; Priabonian of Southern France, Vicentino, Western Transylvania, Georgia; Oligocene of Vicentino (Fig.36).

**Fig.36.** Distribution map of *Cryptoconus sp.*

Subclass **Prosobranchia** Milne-Edwards, 1848

Superorder **Caenogastropoda** Cox, 1959

Order **Neogastropoda** Thiele, 1929

Superfamily **Conoidea** Rafinesque, 1815

Family **Turidae** Swainson, 1840

Subfamily **Borsoniinae**

Genus *Borsonia* Bellardi, 1839

Subgenus *Cordieria* Rouault, 1849

*Borsonia biplicata* (J. Sowerby in Dixon, 1850)

**Material:** 32 specimens (12 well preserved and 20 fragmented)

**Dimensions:** L = 14.36-24.73 mm; W = 12.45-17.85 mm; H = 25.5-39.12 mm

**Description:** Shell is thin, slim, tall, layered. Whirls are prominent in the middle, slowly increasing, and separated by a clear, simple sewing line. Last whirl is large, bulging and ends with a short open channel. Upper part of the whirls is slightly concave and at the bottom is plump, smooth connected. Ornamentation of the shell consists of 7-8 short, wide, raised, transverse ribs. Transverse ribs on the last whirl are lost gradually down. Between each helical rib is inserted in a subtler. Lines of growth are well visible, fine, closely spaced, S-shaped curved (Fig.37).

**Deposits in Bulgaria:** Gorno Ezerovo, Burgas.

**Level in Bulgaria:** Priabonian.

**Total stratigraphic and geographic distribution:** Bartonian of England; Upper Eocene of Northern Germany (Fig.38).

**Fig.37.** *Borsonia biplicata*

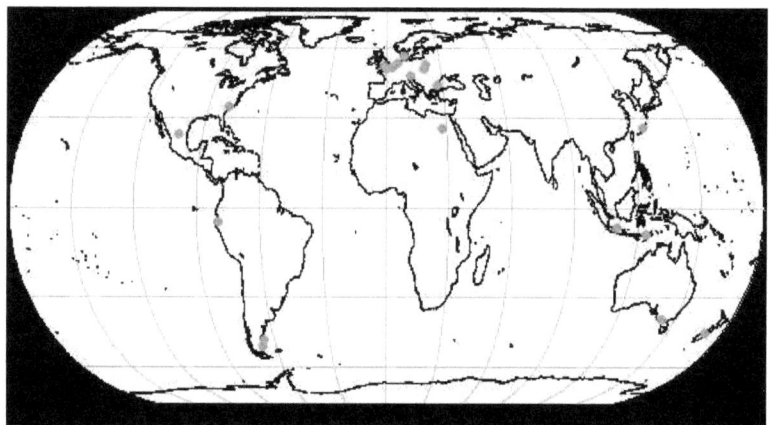

**Fig.38.** Distribution map of *Borsonia sp.*

Subclass **Prosobranchia** Milne-Edwards, 1848

Superorder **Hypsogastropoda** (Ponder and Lindberg, 1997)

Superfamily **Tonnoidea** Suter, 1913

Family **Cassididae** Latreille, 1825

Genus **Cassis** Scopoli, 1777

***Cassis harpaeformis*** (Lamarck, 1803)

**Material:** 2 specimens

**Dimensions:** L = 32.71-39.03 mm; W = 33.41-38.86 mm; H = 44.65-48.74 mm

**Description:** Shell is thick, bi-conical. Whirls are tightly together. Seam is much less pronounced. Last whirl is large, swollen, conical, and narrow at the bottom. Shell surface is decorated with coarse rounded transverse ribs, separated by narrow intercostal, which can be seen clearly only on the last whirl. Ribs end with buds at the top. Near the seam is another spiral line of buds, but smaller. Slightly above the middle of the last whirl there is another spiral line.

**Deposits in Bulgaria**: Kyustendil, Momchilgrad.

**Level in Bulgaria:** Priabonian.

**Total stratigraphic and geographic distribution**: Lutetian of Paris Basin (Fig.39).

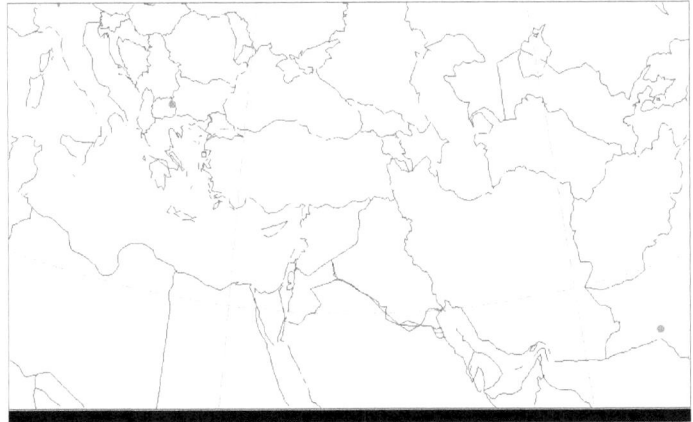

**Fig.39.** Distribution map of *Cassis harpaeformis*

Subclass **Prosobranchia** Milne-Edwards, 1848

Superorder **Hypsogastropoda** (Ponder and Lindberg, 1997)

Superfamily **Tonnoidea** Suter, 1913

Family **Ranellidae** Gray, 1854

Subfamily **Cymatiinae** Iredale, 1913

Genus *Charonia* Gistel, 1848

*Charonia vincenti*

**Material:** 1 specimen

**Dimensions:** L = 9.23 mm; W = 8.71 mm; H = 19.01 mm

**Description:** Shell is high conical, consisting of 5 whirls. They are swollen, angular, separated by a shallow smooth seam. Lower part of the whirls is almost vertical, and the upper is inclined by an angle of 110-120°. Last whirl is significantly large, swollen, and stiff. Ornamentation of the middle whirls consists of 12 thin spiral rounded ribs, equally distant from one another. Between them is inserted a thinner rib. Almost in the middle of the whirls, there is a spiral row of 10-11 buds. They are large, rounded, rarely as small spines. On the last whirl the number of spiral ribs is greater. Radial ribs are quite mild and occur partly on the last whirl.

**Deposits in Bulgaria**: near the town of Somovit, Nikopol, Varna.

**Level in Bulgaria:** Thanetian.

**Total stratigraphic and geographic distribution**: Belgium, Italy, Poland, Denmark, France, Austria (Fig.40).

**Fig.40.** Distribution map of *Charonia sp.*

56

Subclass **Prosobranchia** Milne-Edwards, 1848

Superorder **Hypsogastropoda** (Ponder and Lindberg, 1997)

Superfamily **Tonnoidea** Suter, 1913

Family **Strombidae** Rafinesque, 1815

Genus *Terebellum* Linnaeus, 1758

*Terebellum belemnitoideum*

**Material:** 5 fragmented specimens

**Dimensions:** L = 8.75-17.39 mm; W = 4.91-8.06 mm; H = 27.4-52.46 mm

**Description:** Shell of our specimens is not preserved, which allows us only to see that the suspension is composed of 3 whirls. Last whirl is very large, slightly bulging with broken off bottom. Oral aperture is narrow and elongated.

**Deposits in Bulgaria:** Haskovo.

**Level in Bulgaria:** Priabonian.

**Total stratigraphic and geographic distribution**: In the associations with *Nummulites ramondi* in Asia (Fig.41).

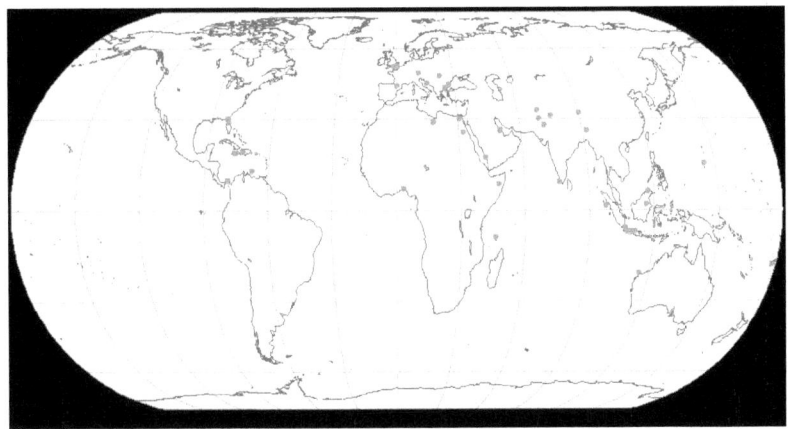

**Fig.41.** Distribution map of *Terebellum sp.*

Subclass **Prosobranchia** Milne-Edwards, 1848

Superorder **Hypsogastropoda** (Ponder and Lindberg, 1997)

Order **Sorbeoconcha** Ponder and Lindberg, 1997

Superfamily **Campaniloidea** Douville, 1904

Family **Campanilidae** Douville, 1904

Genus *Campanile* Fischer, 1884

*Campanile lachesis*

    **Material:** 1 fragmented specimen

    **Dimensions:** L = 8.66 mm; W = 7.23 mm; H = 17.52 mm

    **Description:** Shell is high conical. Initial whirls of our specimen are not preserved. Sewing line is clear, smooth. Each whirl consists of two parts: a lower, barely convex and upper - flat or even slightly concave. Bottom is decorated with large buds and the top is smooth.

    **Deposits in Bulgaria**: Haskovo, Burgas.

    **Level in Bulgaria:** Priabonian.

    **Total stratigraphic and geographic distribution:** Lutetian of Egypt; Bartonian of Vicentino, Dalmatia, Herzegovina; Priabonian of Egypt and Macedonia (Fig.42).

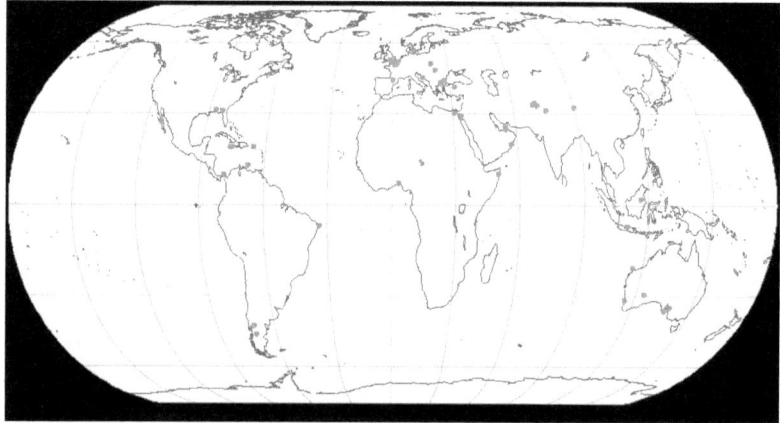

**Fig.42.** Distribution map of *Campanile sp.*

# Phylum MOLLUSCA
## Class Bivalvia

Subclass **Autobranchia** (Groblen, 1894)

Infraclass **Pteriomorphia** (Beurlen, 1944)

Superorder **Ostreiformii** Ferussac, 1822

Order **Ostreida** Ferussac, 1822

Superfamily **Ostreoidea** Rafinesque, 1815

Family **Gryphaeidae** Vialov, 1936

Subfamily **Pycnodonteinae** Stenzel, 1959

Genus *Pycnodonte* Fischer von Waldheim, 1835

*Pycnodonte brongniarti* (Bronn, 1856)

**Material:** 1 left valve

**Description:** Valve has an oval elongated triangular shape. Left ventral side is thick, strongly bulging in the middle. Front end is slightly curved, and the back has a well-developed pinnate knot at the bottom. Beak is great, curved sideways and backwards. Ligament area is large, triangular, with deep triangular fossa, slightly curved forward. Shell is slightly oblique concave under ligament. Surface is decorated with lines of growth, which are more pronounced in the periphery, often lamellar (Fig.43).

**Deposits in Bulgaria:** Priabonian of Haskovo, Krumovgrad; Oligocene of Kyustendil, Kardzhali, Momchilgrad, Svilengrad, Haskovo.

**Level in Bulgaria:** Priabonian-Oligocene

**Total stratigraphic and geographic distribution**: Eocene of Libya, South Afghanistan, Senegal, Pakistan, Czech Republic, Austria; Lutetian of Somalia, Algeria, Tunisia, Southern France, Italy, Northern Transylvania, Turkey, Armenia; Bartonian of France and Italy; Upper Eocene of Sirtika, Cyrenaica, Algeria, Balearic Islands, Northern Italy, the Bavarian Alps, Hungary, South Dagestan; Oligocene of Algeria, Tunisia, Southern France, Liguria, Piedmont, Vicentino, Switzerland, Hungary and Georgia (Fig.44).

**Fig.43.** *Pycnodonte brongniarti*

**Fig.44.** Distribution map of *Pycnodonte sp.*

Subclass **Autobranchia** (Groblen, 1894)

Order **Arcida** (Stoliczka, 1871)

Superfamily **Arcoidea** (Lamarck, 1809)

Family **Arcidae** Lamarck, 1809

Subfamily **Arcinae** Lamarck, 1809

Genus *Barbatia* Gray, 1842

***Barbatia appendiculata sokolovi*** (Kluschnikov, 1958)

**Material:** 1 specimen

**Description:** Shell has an oval rhomboidal shape. Beak is large, released and rolled up, located in the first third of the length of the shell. Ornamentation of the shell consists of 28-29 radial ribs, divided dichotomous and extended downwards from the middle. Lines of the growth are finely lamellar. At the intersection with the radial ribs are obtained rounded buds.

**Deposits in Bulgaria**: Burgas, Haskovo, Aytos.

**Level in Bulgaria**: Priabonian.

**Total stratigraphic and geographic distribution:** Upper Eocene of Ukraine (Fig.45).

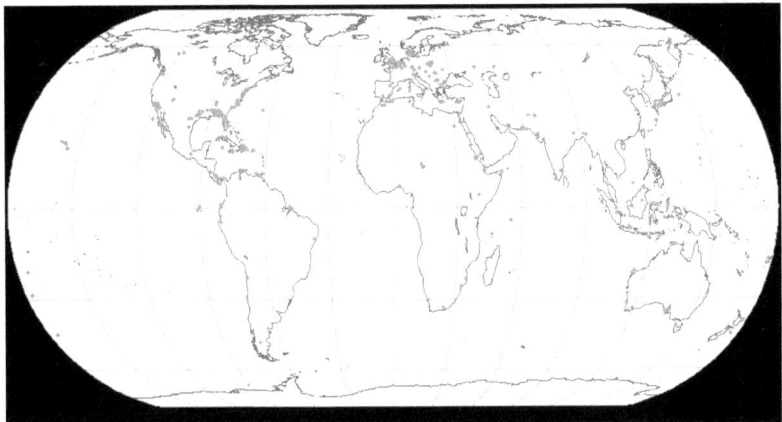

**Fig.45.** Distribution map of *Barbatia sp.*

Subclass **Autobranchia** (Groblen, 1894)

Order **Pectinida** (Gray, 1854)

Suborder **Pectenidina** (Adams and Adams, 1858)

Superfamily **Pectinoidea** Rafinesque, 1815

Family **Pectinidae** Wilkes, 1810

Subfamily **Pectininae** Wilkes, 1810

Genus *Pecten* Müler, 1776

*Pecten* **sp.**

**Material:** 5 fragmented specimens

**Description:** Shell generally is large, thick-oval in a triangular configuration. Ears are large, almost identical, slightly obliquely cut. Right ventral side is usually highly inflated, and the left is flat or concave flat. Both valves are decorated with raised radial ribs which in rare cases are less pronounced. Key is toothless, with a small triangular ligamentous fossa.

**Total stratigraphic and geographic distribution:** The genus is known from the Jurassic to the present (Fig.46).

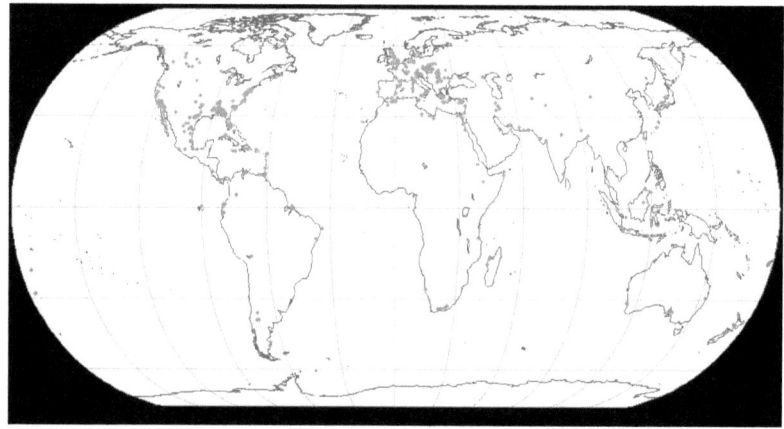

**Fig.46.** Distribution map of *Pecten sp.*

Subclass **Autobranchia** (Groblen, 1894)

Order **Pectinida** (Gray, 1854)

Suborder **Pectinidina** (Adams and Adams, 1858)

Superfamily **Pectinoidea** Rafinesque, 1815

Family **Pectinidae** Wilkes, 1810

Genus *Chlamys* Röding, 1798

*Chlamys rhodopianus* (Bontscheff, 1896)

**Material:** 1 fragmented specimen

**Description**: Shell is medium-sized, rounded, very slightly convex, and almost flat. Front end is a little shorter than the back. Beak is small, pointed, unedited. Ears are large in a straight line. Shell is decorated with 22 radial ribs, separated by deep intercostal. Ribs are covered with thin scales.

**Deposits in Bulgaria**: Kardzhali, Haskovo, Assenovgrad.

**Level in Bulgaria**: Priabonian-Oligocene.

**Total stratigraphic and geographic distribution:** The genus is known from the Triassic to the present (Fig.47).

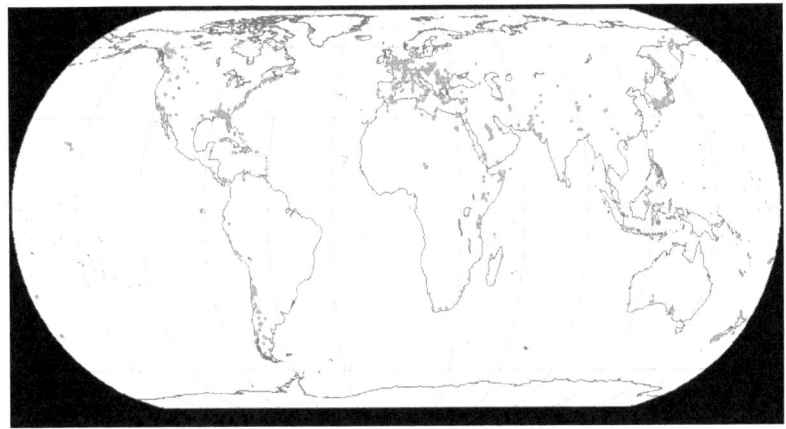

**Fig.47.** Distribution map of *Chlamys sp.*

Subclass **Autobranchia** (Groblen, 1894)

Order **Pectinida** (Gray, 1854)

Suborder **Anomiidina** Gray, 1854

Superfamily **Limoidea** (Rafinesque, 1815)

Family **Limidae** Rafinesque, 1815

Genus *Lima* Mörch, 1853

Subgenus *Ctenoides*

*Lima rara*

**Material:** 4 fragmented specimens

**Description:** Shell is equivalve, elongated oval and moderately tapered shape, slightly convex. Beak is small and sharp. Typical is the sculpture, which consists of 40 radial relief narrow and uniform ribs. Intercostals are wide and deep.

**Deposits in Bulgaria**: Kardzhali

**Level in Bulgaria**: Priabonian.

**Total stratigraphic and geographic distribution:** Lutetian of the Paris basin, the Bavarian Alps; Upper Eocene of North-Western Transylvania (Fig.48).

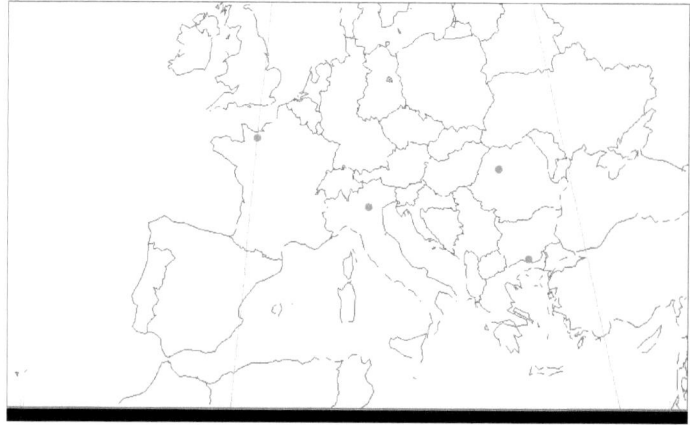

**Fig.48.** Distribution map of *Lima rara*

Subclass **Autobranchia** (Groblen, 1894)

Order **Carditida** Dall, 1889

Superfamily **Crassatelloidea** (Farussac, 1822)

Family **Crassatellidae** Farussac, 1822

Genus *Crassatella* Lamarck, 1799

*Crassatella* **sp.**

Материал: 3 fragmented specimens

**Description:** Shell is medium-sized, triangular, inequilateral, slightly bulging. Beak is small, pointed, slightly curved forward, positioned closer to the front. Decoration consists of concentric ribs.

**Total stratigraphic and geographic distribution:** The genus occurs from the Cretaceous to the present day, cosmopolitan, in warm seas (Fig.49).

**Fig.49.** Distribution map of *Crassatella sp.*

Subclass **Autobranchia** (Groblen, 1894)

Order **Cardiida** Ferussac, 1822

Suborder **Cardiidina** Ferussac, 1822

Genus *Macrosolen*

*Macrosolen* **sp.**


**Material:** 2 specimens

**Description:** Shell is medium-sized, thin, and highly cross-extended with a rounded rectangular shape. Beak is located at the rear end, small, unedited. Ornamentation of the shell consists of many closely spaced fine lines, repeating its outline.

**Total stratigraphic and geographic distribution**: The genus is known from the Paleogene of Europe (Fig.50).

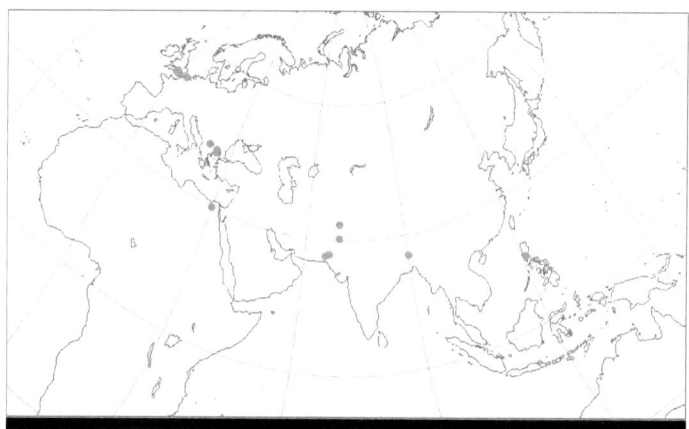

**Fig.50.** Distribution map of *Macrosolen sp.*

Subclass **Autobranchia** (Groblen, 1894)

Order **Carditida** Dall, 1889

Superfamily **Tellinoidea** Blainville, 1814

Family **Tellinidae** Blainville, 1814

Genus *Tellina* Linnaeus, 1758

*Tellina decorata ovalina* Deschayes, 1860

**Material:** 1 specimen

**Description:** Shell is a relatively flat, oval. Beak is small, just released. Surface is cover with fine lines, some of which are more clearly highlighted. Area around the beak is almost smooth.

**Deposits in Bulgaria**: Aytos, Ivaylovgrad.

**Level in Bulgaria**: Priabonian.

**Total stratigraphic and geographic distribution**: Ypresian of the Paris Basin, Bartonian of North Italy (Fig.51).

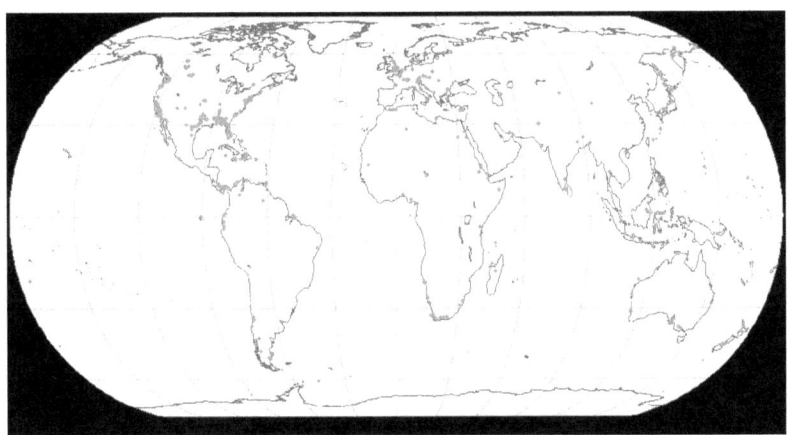

**Fig.51.** Distribution map of *Tellina sp.*

Subclass **Autobranchia** (Groblen, 1894)

Order **Carditida** Dall, 1889

Superfamily **Cardioidea** Lamarck, 1809

Family **Cardiidae** Lamarck, 1809

Subfamily **Orthocardiinae** Schneider, 2002

Genus *Cardium* Tremlett, 1950

*Cardium gratum*

**Material:** 1 fragmented specimen

**Description:** Shell is rounded, almost equilateral, and strongly but evenly convex. Beak is small, issued rolled forward. Front end is a little shorter than the back. Decoration of the shell comprises 38-40 radial ribs, separated by a somewhat narrower deep intercostal. Ribs are flat, smooth.

**Deposits in Bulgaria:** Aitos, Kyustendil, Kardzhali.

**Level in Bulgaria**: Priabonian.

**Total stratigraphic and geographic distribution**: Lutetian of the Paris basin, the Bavarian Alps, Egypt; Bartonian of the Paris Basin; Priabonian of Hungary, Northern Italy, Southern France, Egypt, Northern Germany (Fig.52).

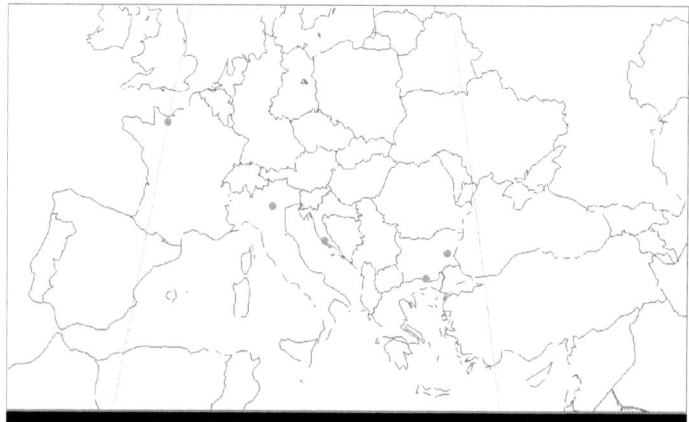

**Fig.52.** Distribution map of *Cardium gratum*

68

Subclass **Autobranchia** (Groblen, 1894)

Order **Carditida** Dall, 1889

Superfamily **Veneroidea** (Rafinesque, 1815)

Family **Veneridae** Rafinesque, 1815

Subfamily **Pitarinae** Stewart, 1930

Genus *Pitar* Römer, 1857

*Pitar bonnetensis* (Boussac, 1911)

**Material:** 3 fragmented specimens

**Description:** Form of the oval shell is triangular, inequilateral. Beak is small, tapered and rolled forward, located in the first third of the length. Front end is short, rounded. Back end is longer, curved. Decoration of the shell consists of fine concentric ribs which are densely located.

**Deposits in Bulgaria**: Ivaylovgrad; Krumovgrad, Tarnovo.

**Level in Bulgaria**: Priabonian.

**Total stratigraphic and geographic distribution**: Lower Priabonian of Southern France, Bavarian Alps, Georgia (Fig.53).

**Fig.53.** Distribution map of *Pitar sp.*

Subclass **Autobranchia** (Groblen, 1894)

Order **Carditida** Dall, 1889

Superfamily **Veneroidea** (Rafinesque, 1815)

Family **Veneridae** Rafinesque, 1815

Subfamily **Pitarinae** Stewart, 1930

Genus *Pitar* Römer, 1857

*Pitar villanovae* (Deshayes in Studer, 1853)

**Material:** 1 fragmented specimen

**Description:** Shell is inequilateral, strongly bulging around the beak, with oval triangular shape. Beak is a small, pointed, slightly curved downward and forward, situated closer to the front end. Shell surface is covered with coarse broad concentric ribs separated by narrow and shallow intercostal. In the middle of the ribs are the most rude, and at the end are more subtle.

**Deposits in Bulgaria**: Ivaylovgrad, Aytos, Kardzhali.

**Level in Bulgaria**: Priabonian.

**Total stratigraphic and geographic distribution**: Bartonian of Switzerland, Hungary, Egypt; Priabonian of South France, Western Alps, Vicentino, Egypt; Oligocene of Romania, Northern Italy, Armenian (Fig.53).

## VII. PALEO-ECOLOGICAL CHARACTERISTIC OF THE STUDIED REGION

*GEOLOGICAL TIME*

Based on the identified mollusc fossils, the age of the studied fossil trove could be related to the Late Eocene – Priabonian, according to data from the reference sources (Karagyuleva, 1964; Abbott & Dance, 1986; Harzhauser & Mandic, 2001; Harzhauser 2004; Kiessling, 2004; Hendy et al., 2009).

Geological distribution of 23 of all identified 36 species (63.8%) is only in Priabonian, as follows: 17 gastropod taxa - *Tympanotonus trochleare diaboli, Chondrocerithium filigrana, Sycostoma bulbiforme, Lyria decora, Rimella multiplicata, Ficus helveticus, Globularia grossa, Globularia patula, Globularia sigaretina, Globularia vapincana, Cepatia cepacea, Euspira achatensis, Borsonia biplicata, Terebellum belemnitoideum, Campanile lachesis, Cassis harpaeformis, Burtinella spirulaea;* and 6 bivalve taxa - *Lima rara, Cardium gratum, Pitar bonnetensis, Pitar villanovae, Tellina decorate ovalina, Barbatia appendiculata sokolovi.*

Nine taxa (25%) are with wider geological range (Eocene-Oligocene): 7 gastropods - *Velates perversus, Cirsotrema bourdoty, Canarium auriculatum, Amaurellina angustata, Globularia crassatina, Tectus lucasianus, Cryptoconus filosus;* and 2 bivalves - *Pycnodonte brongniarti* and *Chlamys rhodopianus.*

Most of described gastropods are stratigraphically and geographically widespread species. In Bulgaria, they are found in the regions of Burgas, Chirpan, Haskovo, Kardzhali, Krumovgrad, Aytos, Tarnovo, and Ivaylovgrad. They are also reported from the Paleocene to the Early Eocene of Europe.

Moluscs fauna of France (Loire-Paris Basin), Italy (Venetian & Piedmont Basin) and other large European basins during the Eocene is very similar or even identical to our findings. Environmental conditions should have been relatively close. Bivalves and gastropods from shallow seas are a good basis for paleobiogeografical correlations and also to some extent for stratigraphic comparisons (Piccoli, 1984; Piccolo et al., 1986; Amitrov, 1994).

71

Gastropods of the genus *Tympanotonus* are regarded as facial indicators of lagoon to the rocky conditions and oligo/meso salinity (Baldi, 1973; Barthelt, 1989). Species *Tympanotonus margaritaceus* is widespread in sub littoral of the coastal marshes (Harzhauser, 2004) and also has been reported for lagoons and brackish waters in riverine-estuary facies (Harzhauser & Mandic, 2001).

Gastropods of the genus *Globularia* have been reported as indicators of typical marine conditions in the East Mediterranean (Mesohellenic Basin, Greece) (Harzhauser, 2004). This type of gastropod fauna reflects the increase in salinity and a gradual transition from littoral environment to sub littoral one.

During the Eocene the marine mollusc fauna in Europe was very similar between the different sea basins, as for example the Loire-Paris Basin in France, and the Venetian and Piedmont Basin in Italy. The environmental conditions in the shallow sea areas were similar and the gastropods and mussels are a good base for paleobiogeographic correlations and stratigraphic comparisons (Piccoli, 1984; Piccolo et al., 1986; Amitrov, 1994).

Bivalves live in different types of water basins - from high to the low salinity. Most of identified taxa from the studied fossil trove could not tolerate even slight fluctuations in abiotic factors. This fact makes them reliable indicators of environmental conditions existing in Paleogene marine basin in the Eastern Rhodopes (Temelkov & Osman, 2008).

Species of the families Pitar, Cardium, Macrosolen, Lima, Chlamys and Crassatella are marine benthic inhabitants on soft (sandy) substrate.

One of studied fossil species, *Pycnodonte bongniarti*, is referred to saline and brackish conditions.

Genus *Barbatia* is distributed in various environment – lagoon, coastal, inner shelf and outer shelf, oceanic, brackish, and freshwater.

Most of the taxa are mesotermophyls and indicate moderately warm and humid climate in the region (Temelkov & Cholakov, 1996).

Structure: It is well known that the majority of plant and animal species in the ancient biocenosis are not preserved in the fossil state. Thus, in the form of fossil remains, in most cases, we reached only about 30% of all taxa. So, paleoecological reconstructions are not able to represent the whole paleobiocenosis and can only be approached to some extent. Although the fossil record is incomplete, it still provides a useful survey of the history of life because of the vast amounts of time represented within the rock record.

Based on the identified mollusc fauna, it can be point out some ecological properties of the ancient gastropod and bivalve complexes.

Gastropods strongly predominated with 618 specimens while the mussels were rare, and were less than 6% from all collected molluscs. We described 392 gastropod fossils relative to 17 genera, while the remaining 226 specimens we were unable to identify because of the poor preservation. Genus Globularia was dominant in the gastropod complex both of quantity (31.1%) and of a number of taxa with which it is presented (29.4%), followed by Tympanotonus sp. (13%) and Cirsotrema sp. (11%) (Fig.54).

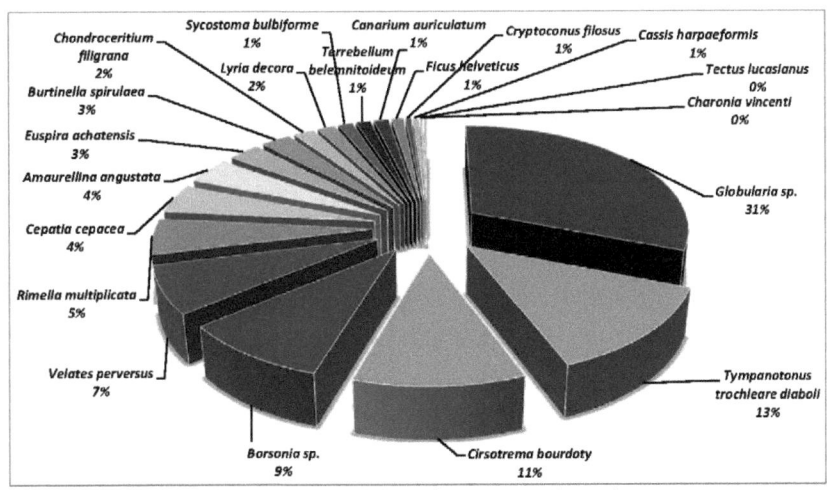

Fig.54. Structure of the ancient gastropod complex

From the collected 36 bivalve specimens we described 27 fossils relative to 10 genera, while the remaining 9 specimens we were unable to identify because of the poor preservation. Pecten sp. was presented with 5 individuals, and Lima sp., Chlamys sp. and Pitar sp. – with 4 individuals (Fig.55).

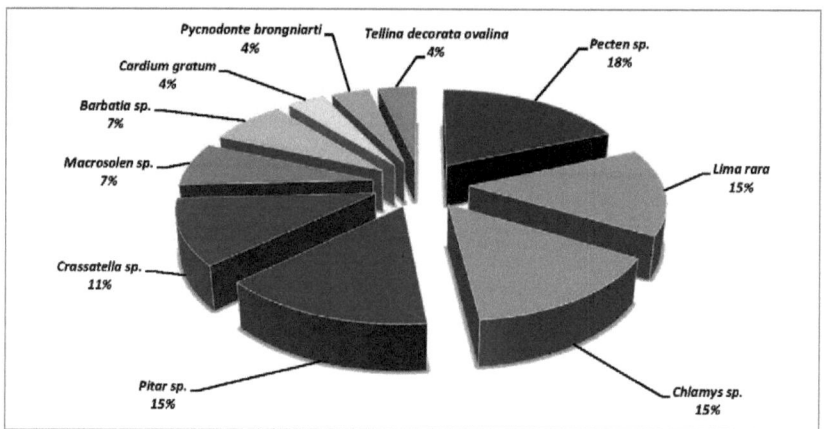

Fig.55. Structure of the ancient bivalve complex

Diversity: Diversity indices calculated for gastropod and bivalve complexes are presented in Table 3. Index of dominance (Dominance_D) is close to the minimum, slightly lower in bivalves. This means that the number of individuals with whom each taxon are presented in the complex, is similar, there are no taxa with a predominant amount of specimens.

High values of the index of uniformity (Equitability, J) and Simpson index (Simpson = 1-D) confirm the uniformity of the identified taxa of gastropods and bivalves. Uniformity is higher in bivalves than that of gastropods, which means that in the complex these taxa are more evenly distributed.

Margalef's index provides a more detailed assessment of the species richness of the community because it can accept values greater than 1 (not only between 0 and 1). Thus facilitates biodiversity comparisons between communities of different habitats and regions. In our study, this index confirms the higher diversity of gastropods complex as compared with that of the bivalves.

74

Index of Berger-Parker gives data of the ratio between the number of individuals of the most abundant taxon and the total number of individuals. As seen on Fig.54 and Fig.55, genus *Globularia* significantly prevailed from gastropods, and in bivalve complex there are four genera with almost equal quantity.

**Table 3.** Diversity indices on the base of identified gastropod and bivalve taxa

| Index | Gastropoda | Bivalvia |
|---|---|---|
| Taxa | 21 | 10 |
| Individuals | 392 | 27 |
| Dominance_D | 0.1474 | 0.1276 |
| Simpson (1-D) | 0.8526 | 0.8724 |
| Shannon (H) | 2.339 | 2.157 |
| Simpson Evenness (E) | 0.4936 | 0.8644 |
| Margalef | 3.349 | 2.731 |
| Equitability (J) | 0.7681 | 0.9367 |
| Berger-Parker | 0.3112 | 0.1852 |

**Ecological groups:** Identified fossil species of the two mollusc classes from the studied paleobiocenosis could be related to the following ecological groups of organisms:

**1) On the base of their feeding behaviour:**

- Suspension-feeders (10 taxa, bivalves): *Pycnodonte bongniarti, Lima rara, Pitar bonnetensis, Pitar villanovae, Pecten sp., Chlamys rhodopianus, Barbatia appendiculata sokolovi, Crassatella sp., Macrosolen sp., Cardium sp.*

- Deposit-feeder (1 species, bivalve): *Tellina decorata ovalina*

- Carnivores (11 taxa, gastropods): *Ficus helveticus, Charonia vincenti, Cepatia cepacea, Euspira achatensis, Cirsotrema bourdoty, Borsonia biplicata, Lyria decora, Sycostoma bulbiforme, Globularia sigaretina, Cassis harpaeformis, Cryptoconus filosus*

75

- Grazers (8 taxa, gastropods): *Tympanotonus trochleare diaboli, Campanile lachesis, Amaurellina angustata, Globularia vapincana, Globularia grossa, Globularia patula, Globularia crassatina, Tectus lucasianus*

- Omnivore-Grazers (4 taxa, gastropods): *Rimella multiplicata, Canarium auriculatum, Terebellum belemnitoideum, Velates perversus*

- Herbivore (1 species, gastropod): *Chondroceritium filigrana*

**2) On the base of their life habit:**

- Epifaunal: 22 taxa of gastropods and 5 taxa of bivalves

- Infaunal: 3 taxa of gastropods and 5 taxa of bivalves

## VIII. CONCLUSION

As a synopsis of this study, it could be concluded that the area of the Perunika village, where is located the fossil deposit, was a part of a large marine basin that existed during the Paleogene in the Eastern Rhodopes. Typically for the entire Paleogene marine basin was the labile seabed because of frequent tectonic movements, followed by lowering the bottom. Identified invertebrates and associated biota suggest warm (18-22°C) and shallow sea (up to 200 m), with oligo to mezo salinity, probably part of a sublittoral zone. Bottom of the studied section of the sea was mostly rocky, and in shallow areas - sandy. Brightness and hydrodynamics of water were favorable for the development of reef corals. Paleoecological characteristics showed that life in the Paleogene sea basin in the area of Perunika village was by a rich diversity of species, complex trophic structure, multiple biotic interactions and dynamic equilibrium. Studied fossil trove is remarkable due to the abundance and diversity of fossil marine invertebrates.

## IX.  REFERENCES

1.    Abbott R., S. Dance. 1986. Compendium of Sea Shells, E. P. Dutton, pp. 412.

2.    Aberhan M., J. Alroy, F.T. Fursich, W. Kiessling, M. Kosnik, J. Madin, M. Patzkowsky and P. Wagner. 2004. Ecological attributes of marine invertebrates. Accessed on 05 October 2013. <URL: http://paleodb.org/bridge.pl?a=displayReference&reference_no=9941>

3.    Amitrov O. 1984. Changes in composition of the Gastropods in the western Eurasian seas at the Eocene-Oligocene Boundary. – Paleontological Journal, 28: 19-30.

4.    Baldi T. 1973. Mollusc fauna of Hungarian upper Oligocene (Egerian). Akademiai Kiado, Budapest, pp. 511.

5.    Barthelt D. 1989. Faziesanalyse und Untersuchungen der Sedimentations mechanismen in der Unteren Brackwasser-Molasse Oberbayerns. Munchner Geowissenschaftliche

6.    Beu A. G., P.A. Maxwell, R.C. Brazier. 1990. Cenozoic Mollusca of New Zealand. *New Zealand Geological Survey Paleontological Bulletin,* **58**.

7.    Boshev S., Zafirov S., Krastev B. 1966. Geology of Bulgaria with paleonthology and historical geology. Tehnika Publ., Sofia, 384 p. (In Bulgarian).

8.    Boyanov I., B. Mavrudchiev, I. Vaptzarov. 1963. Notes on the structural peculiarities of the part of East Rhodopes. Institut of Geology, BAS, 12. (In Bulgarian)

9.    Boyanov I. 1971. Paleogenic and Neogenic depressions on the eastern part of the East-Rhodopean massif. Technika, Sofia, 127-135. (In Bulgarian)

10.    Boyanov I., A. Goranov. 2001. Late Alpine (Paleogene) superimposed depressions in parts of Southeast Bulgaria. Geologia Balkanica, 31, 3-36. (In Bulgarian)

11.    Bunje P. 2003. The Mollusca. University of California Museum of Paleontology. <URL: http://www.ucmp.berkeley.edu/taxa/inverts/mollusca/mollusca.php>. Accessed 26 September 2013.

12.    Eames F. E.. 1951. A Contribution to the study of the Eocene in Western Pakistan and Western India: B. Description of the Lamellibranchia from standard sections in the Rakni

Nala and Zinda Pir areas of the Western Punjab and in the Kohat District. Philosophical Transactions of the Royal Society of London, Series B, 235: 311-482.

13.  Ivanov R. 1960. The magmatism in East-Rhodopean Paleogene depression. I.Geology. Proceedings on Geology of Bulgaria, 1: 312-387. (In Bulgarian)

14.  Ivanov R. 1961. Stratigraphy and structure to the metamorphic complexes in East Rhodopes. Proceedings on Geology of Bulgaria, 2: 69-119. (In Bulgarian)

15.  Gastaldo R., S. Savrda, & Lewis. 1996. Deciphering Earth History: A Laboratory Manual with Internet Exercises. Contemporary Publishing Company of Raleigh, Inc. ISBN 0-89892-139-2.

16.  Gekker, R. F. 1957. Vvedenie v Paleoekologiyu, Moscow, pp. 83.

17.  Georgiev V. 2002. Origin and initial evolution of the Momchilgrad volcanotectonic depression (Eastern Rhodopes). Review of the Bulgarian Geological Society, 63, 1-3: 99-106. (In Bulgarian)

18.  Goranov A. 1960. Litology of the Paleogenic sediments in the part of East Rhodopes. Proceedings on Geology of Bulgaria, 1: 258-310. (In Bulgarian)

19.  Hammer Ø., D. Harper, P. Ryan. 2001. PAST: Paleontological Statistics Software Package for Education and Data Analysis. Palaeontologia Electronica, 4-1: 9.

20.  Harzhauser M., O. Mandic. 2001. Late Oligocene gastropods and bivalves from the Lower and Upper Austrian Molasse Basin. In: Piller WE, Rasser MW, editors, Paleogene of the Eastern Alps. Verlag der Osterreichischen Akademie der Wissenschaften, Vienna, pp. 671-795.

21.  Harzhauser M. 2004. Oligocene gastropod faunas of the eastern Mediterranean (Mesohellenic Trough/Greece and Esfahan-Sirijan Basin/Central Iran). Cour Forschung Sencken 248: 93-181.

22.  Hendy A., M. Aberhan, J. Alroy, M. Clapham, W. Kiessling, A. Lin, M. LaFlamme. 2009. Unpublished ecological data in support of GSA 2009 abstract: A 600 million year record of ecological diversification. Accessed on 05 October 2013. <URL: http://paleodb.org/bridge.pl?a=displayReference&reference_no=29272>.

23. International Commission on Stratigraphy. Accessed on 05 October 2013. <URL: http://www.stratigraphy.org>.

24. Karagyuleva Y. 1964. Les fossiles de Bulgarie, VI a, Paleogene Mollusca. BAS, Sofia, pp. 164

25. Kiessling W. 2004. Ecology opinions. Accessed on 05 October 2013. <URL:http://paleodb.org/bridge.pl?a=displayReference&reference_no=9940>.

26. Mikkelsen P., R. Bieler. 2008. Seashells of Southern Florida. Living marine mollusks of the Florida Keys and regions, pp. 503

27. Petrova S., E. Mehmed, I. Mollov, D. Georgiev, I. Velcheva. 2012. Molluscs (Mollusca: Gastropoda, Bivalvia) from the Upper Eocene of Perunika village (East Rhodopes, Bulgaria) – Preliminary results. Acta Zool Bulgar Suppl. 4: 237-240.

28. Picolli G. 1984. Cenozoic molluscan associations of Mediterranean and South-East Asia: a comparison. Memorie di Scienze Geologische, 36: 499-521.

29. Picolli G., S. Sartori, A. Franchino. 1986. Mathematical model of the migration of Cenozoic benthic Molluscs in the Tethyan belts. Memorie di Scienze Geologische, 38: 207-244.

30. Sanchez Roig M. 1926. Los Equiodermos fosiles de Cuba.Contribucion a la paleontologia Cubana, pp. 179.

31. Sapoundjieva V. 1964. Les fossiles de Bulgarie, VI b, Paleogene Echinoidea. BAS, Sofia.

32. Steuber T. 2002. Web Catalogue of the Hippuritoidea (rudist bivalves). Accessed on 05 October 2013. <URL: http://www.ruhr-uni-bochum.de/sediment/rudinet/index.htm>

33. Temelkov B., N. Cholakov. 1996. Donnees paleoecologiques sur les bivalves priabonien des versants nord du Rhodope (Bulgarie). Travail Scientific des Université de Plovdiv, Animalia, 32, 6:23-33. (In Bulgarian)

34. Temelkov B., M. Osman. 2008. Paleoecological investigation of the palaeogenous invertebrates in the Kardzali region. In: Velcheva I., A. Tsekov (2008), Proceedings of the Anniversary Scientific Conference of Ecology, Plovdiv.

35.   Todd J. A. 2001. Bivalve Life Habits. Accessed on 05 October 2013. <URL: http://paleodb.org/bridge.pl?a=displayReference&reference_no=30184>

36.   Waller T.R. 1998. Origin of the molluscan class Bivalvia and a phylogeny of major groups. Pp. 1-45 in P.A. Johnston and J.W. Haggart (eds.), An Eon of Evolution; Paleobiological Studies Honoring Norman D. Newell. University of Calgary Press.

37.   Yordanov M. 1962. Historical geology and geology of Bulgaria. Nauka I Izkustvo Publ., Sofia, 235 p. (In Bulgarian)

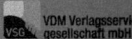

Printed by Books on Demand GmbH, Norderstedt / Germany